ALGAL BIOFUELS

ALGAL BIOFUELS

Editor

Leonel Pereira

MARE—Marine and Environmental Sciences Centre
and
IMAR—Institute of Marine Research
Department of Life Sciences
University of Coimbra
Coimbra
Portugal

CRC Press
Taylor & Francis Group
Boca Raton London New York

CRC Press is an imprint of the
Taylor & Francis Group, an **informa** business

A SCIENCE PUBLISHERS BOOK

Cover credits:
· Top right image: A photograph from 'A4F—Algae for Future, S.A., Lisboa, Portugal'—reproduced with permission of Drs. Vitor V. Vieira, Luis T. Guerra, Joana F. Lapa, Diana B. Fonseca, Luis F. Costa and Edgar T. Santos (authors of Chapter 1)
· Top left, bottom left images: Photographs of the Photobioreactors used in the algae plant, Algafarm, Pataias, Portugal—reproduced with permission of Prof. Leonel Pereira (editor of the book).

CRC Press
Taylor & Francis Group
6000 Broken Sound Parkway NW, Suite 300
Boca Raton, FL 33487-2742

First issued in paperback 2021

© 2017 by Taylor & Francis Group, LLC
CRC Press is an imprint of Taylor & Francis Group, an Informa business

No claim to original U.S. Government works

Version Date: 20170516

ISBN-13: 978-0-367-78210-8 (pbk)
ISBN-13: 978-1-4987-5231-2 (hbk)

Visit the Taylor & Francis Web site at
http://www.taylorandfrancis.com

and the CRC Press Web site at
http://www.crcpress.com

Preface

The use of algae (micro and macroalgae) for biofuels production is expected to play an important role in securing an alternative energy supply in the next decades. The potential of algae as fuel of the future is a very important topic given the shortage of fossil fuel reserves and its environmental impact. Algae are presented as a viable alternative because the production of algae for fuel should not compete with the food, preventing the increase in food prices, and is the third generation of biofuels. Society's pressure is already high enough for academia and industry to find alternatives to fossil fuels, but the production of biofuels from algae, in an economically viable way, maybe will only be possible when fossil fuel sources are nearly exhausted.

A quick summary of each chapter is presented below:

Chapter 1—Microalgae are extremely competitive with other plant crops and have an enormous potential to be used as feedstock for biofuels and biorefinery products. Microalgae are highly efficient and can be integrated with carbon fixation from waste gas streams. Moreover, they do not compete with the food and feed sector since they can grow in non-arable land, can be produced in a continuous process and are not seasonal. However, it is well accepted that microalgae industrial production must be optimized in order to be economical competitive with higher plants since large quantities of growth medium, water and energy are required.

Chapter 2—Microalgae is one of the most promising source of biofuels production. However, the production costs are still high, due to the increase of fertilizers' cost. Using wastewater as culture medium not only results in the decrease of production costs, but can also constitute an efficient tertiary treatment of these effluents. Thus, this chapter discusses current developments and trends concerning the usage of wastewater resources as a way of growing microalgae for biofuel production purposes. Besides presenting the typical characteristics and particularities of different wastewater streams, methodologies to improve of microalgae productivities are also given. Main issues regarding wastewater phycoremediation are also assessed and linked with the viability of subsequent biofuel production.

Case studies concerning different wastewater sources are presented and key research topics are suggested.

Chapter 3—Microalgae have been suggested as good candidates for fuel production because of their higher photosynthetic property, efficiency, higher biomass production, and faster growth compared to other energy crops. Algae contain protein, carbohydrates and lipids. Lipids can be processed to biodiesel, carbohydrates to be ethanol and H_2, and proteins as raw material for biofertilizer. There are many ways to convert the oil and fats into biodiesel, namely transesterification, esterification, blending, micro emulsion and pyrolysis, but transesterification and esterification are the most commonly used methods.

Other important factor in biodiesel production is fatty acids (FA) type, and its amount. There are three main type of the FA that can be present in a triglyceride, i.e., saturated, mono-unsaturated and poly-unsaturated with two or three double bonds. The cetane number, kinematic viscosity, density and heating value of biodiesel can be predicted from the FA composition.

Chapter 4—Macroalgae (also called seaweeds) have gained attention in the last years as feedstock for the production of fuels and chemicals due to the advantages they show with respect to traditional terrestrial feedstocks for biorefinery: (i) higher productivity of cultivation (biomass produced per unit of surface) than terrestrial crops, (ii) no competition for arable land, (iii) lower fresh water consumption during cultivation and (iv) no requirement for fertilizers.

The biochemical composition of macroalgae, containing sugars, proteins and minerals as main components, make them suitable to be used in biorefineries for the production of fuels, chemicals and other products, valorizing all components in the biomass. In this chapter an overview of the use of macroalgae for biofuel production from a biorefinery perspective is given. Types of seaweeds, their composition, processes for isolation of carbohydrates for biofuel production, and biochemical and thermochemical conversion processes to biofuels, including biomethane, is given. Attention is paid to developments in large-scale cultivation of macroalgae and to the variation in composition between the different species. In addition, uses of proteins in macroalgae will be discussed as part of the wider picture of using seaweeds as feedstock for the future biobased economy.

Chapter 5—The use of algal biomass for bio fuels production has many advantages, namely: algal bio-mass can be produced all over the year and have a rapid growth capability, it grows in aqueous media, but it requires less water than terrestrial crops, it can be grown in brackish water on non-arable land, algae cultivation does not need herbicides or pesticides application and the nutrients for algae growth (mainly nitrogen and phosphorus) can be obtained from wastewater, thus as the algae grow, water effluent from agro-industrial sectors are treated. Algae growth does not compete with food production and it improves air quality due to CO_2 bio-fixation, as 1 kg of dry algal biomass uses around 1.83 kg of CO_2. Many species of microalgae have oil content between 20 and 50% dry weight and by changing the growth conditions the oil yield may increase significantly.

The thermochemical processes developed for biomass energetic valorization may be also used for algal biomass, having in mind the specificities of this type of biomass.

Thermochemical processes are usually divided into dry or conventional and wet or new hydrothermal processes that operate under sub or supercritical conditions.

Chapter 6—The renewable bioethanol fuel used at present originates from agricultural and forest biomass. Brazil is the biggest ethanol producer originating in fermentation of non-crystallizing sugarcane molasses waste, being that about 55% of Brazilian sugarcane is converted to ethanol. Canada and the USA convert part of its cereal production to ethanol, with a value of about 40% corn converted to ethanol in the USA. However, these sources can only produce part of the required amount of biofuels and make strong competition with food production resources and land use and depletion. Until 2050, there is a need to increase food production by 70% relative to the level of 2005 to satisfy the demand of an increasing population need. For the same duration, the increase in land available for agriculture is expected to rise only by about 5%. Finding others sources of energy is an urgent goal and more research in alternatives is needed. The lignocellulosic materials, mostly considered waste materials derived from agricultural and forest activities do not compete with food and land use and their conversion is very promising. In fact, a lot of effort was made in the last decade to try to achieve the technological means to make its conversion into biofuels. However, technology demonstrated that is difficult to convert the lignocellulosic carbohydrates in biofuels, mainly due its intricate and strong structure where the high crystallinity of cellulose and the presence of lignin are two main problems.

Leonel Pereira

Contents

CHAPTER 1

Microalgae Biotechnologies
Possible Frameworks from Biofuel to Biobased Products

Vitor V. Vieira,* Luis T. Guerra, Joana F. Lapa, Diana B. Fonseca, Luis F. Costa and *Edgar T. Santos*

1. Introduction

Microalgae are extremely competitive with other plant crops and have an enormous potential to be used as a feedstock for biofuels and biorefinery products. They are highly efficient and can be integrated with carbon fixation from waste gas streams. Moreover, they do not compete with the food and feed sector since they can grow in non-arable land, can be produced in a continuous process and are not seasonal. However, it is well accepted that microalgae industrial production must be optimized in order to be economically competitive with superior plants since high quantities of growth medium, water and energy are required.

The biorefinery concept aims to convert microalgae biomass in several low and high value products, e.g., biofuels, bioalcohols, biochemicals, complex chemicals and bioproducts such as omega-3 fatty acids, pigments, etc. The full value-chain of the biomass is the critical step since it is important to involve several market sectors. Also, the biorefinery as a whole is still in proof of concept stage and market implementation remains a challenge. Therefore, a variety of technologies are being developed and improved; from harvesting and feedstock pretreatments to post-processing techniques,

A4F – Algae for Future, Estrada do Paço do Lumiar, Campus do Lumiar, Ed. E – R/C, 1649-038 Lisbon, Portugal.
* Corresponding author: vvv@a4f.pt

such as biological and thermochemical procedures as well as system integration and evaluation. Harvesting processes such as sedimentation, flocculation, and centrifugation are currently being used and improved, but other methods are also being accessed, e.g., selective membranes and nanoparticles.

Through biological processes, it is possible to transform the biomass into carbon chains with two to six carbons, using fermentation and anaerobic digestion techniques. In order to estimate and access the theoretical sustainability impact of the global process, considering various aspects, i.e., technical data/characteristics, environmental, economic and social criteria, it is necessary to perform a consistent Life Cycle Assessment. Through the combination of different methodologies and processing systems, several scenarios can be conjectured, enabling a sustainable management of the environmental impacts of future essays.

2. Possible Frameworks

The context and options for microalgae application are a result of a specific framework. These options depend on present and emerging needs plus accumulated evolution achievements (Fig. 1.1). To understand the possible future pathways of the microalgae sector, it is necessary to study the historical evolution, the existing knowledge and experience, and combine that with market future needs.

At present **we are going through a paradigm change**:

"There are several emerging market needs that can be best fulfilled by a new agriculture crop: microalgae."

As a result of accumulated knowledge, at this moment, there is 'critical mass' that makes it possible to have this new agriculture crop in a transition from research to commercial scale. The author has been monitoring in detail the microalgae sector since 1989. More than 300 events (academic,

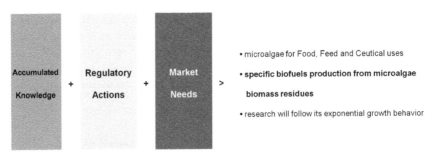

Figure 1.1. Food, Feed, and Ceuticals will overcome biofuel applications.

industry related, and political or regulatory) involving microalgae were evaluated along the last 310 years (since 1703, see Table 1.1), as well as a list of more than 900 companies currently in operation that have been followed on a regular basis and more than 350 that went out of business. Three simultaneous inputs were spotted as the change-making factors:

(1) **Accumulated knowledge:** The first description of microalgae was reported 310 years ago. There was a slow progress in early times, then a linear growth in the 20th century—and an exponential during the last ten years. This growth happened both in the number of persons working with microalgae at the research level and in the number of start-up companies. Following the oil crisis in the seventies, in the US, the Carter administration through the Department of Energy under the Office of Fuels Development, funded the Aquatic Species Program, where the National Renewable Energy Laboratory scientists isolated around 3,000 algae species over a period of nearly 20 years from 1978 to 1996. The first known company was *Chlorella Industry* in Japan in 1964. Forty years later, in 2003, there were less than 100 companies worldwide involved in the continuous production of microalgae, while ten years later, in 2013, there were more than 1,000. Less than 20% of companies existing in 2003 still exist now. The number of research groups and research organizations with projects related to microalgae went from less than 300 worldwide to more than 5,000 in the same period.

Microalgae are extremely interesting as a new crop but not as a 'miracle-making crop'. Several review papers, published in 2006 and 2007, brought together and reviewed most of the scientific and technological information. A new idea came up: if certain microalgae can have up to 70% oil; double every day and have high productivities—then, 'microalgae' can produce 136,900 L/ha (Chisty 2007). It was also concluded that "microalgae appear to be the only source of biodiesel that has the potential to completely displace fossil diesel. Unlike other oil crops, microalgae grow extremely rapidly and many are exceedingly rich in oil" (this reference had more than 3,000 citations in peer-reviewed papers). Obviously this syllogism, which was cited and repeated through the years, drives from an incorrect association of metabolic pathways, since there is no organism able to use the absorbed energy and maximizing simultaneously two different function objectives: energy accumulation in the form of lipids and cell division. This syllogism breaks the principle of energy conservation, as productivities depend also on energy balance. A similar reasoning was made about microalgae for CO_2 'sequestration' assuming impossible levels that forgot to consider the mass conservation principle: microalgae cannot fix more carbon than their photosynthetic theoretical limit.

(2) **Regulatory factors:** 'Microalgae for Fuels' was pushed by political reasons & regulatory factors, and pulled by relevant science & technology to a level of worldwide awareness and commercial interest. These were the triggering factors for the recent interest on microalgae. The US Energy Policy Act of 1992 directed a demand for more studies to be done on biofuels, and gave some guidance to federal programs for the increased implementation of biofuels. In Europe, a proposal to develop an EU-15 Biofuel Directive was launched during 2001. The peak in oil prices in 2008 boosted new interest in algal fuel worldwide. Research programs were initiated to investigate the different processes required to produce algal fuel.

In December 2008, the EU struck a deal to satisfy 10% of its transport fuel needs from renewable sources, including biofuels, hydrogen, and green electricity, as a part of negotiations on its energy and climate package. On 1st July 2011, the American Society for Testing of Materials officially approved the use of algae—and other sustainably-derived biofuels, in commercial and military aircrafts. Under EU Directive 2009/28, the European Commission decided to include algae biomass in the first place of the list of substances which will count four times their energy content towards the overall 10% EU target for renewable fuels in transport. In addition, the EU Emissions Trading System (ETS), the largest multinational, multi-sector greenhouse gas emission trading system in the world, is intending to support EU meeting its 20% emissions reduction target by 2020. The EU ETS covers more than 11,000 energy-intensive industrial installations throughout Europe, such as power stations, refineries, large manufacturing plants, and was expanded to the aviation industry on 1st January 2012.

(3) **Market opportunities:** There is a current need for alternative PUFA oils to start replacing fish oils and fishmeal. Since around 2005, aquaculture feeds has continued its strong annual growth of around 7%, but the volumes of fishmeal used in aquaculture have remained steady at around 3.2 million tons, and those of fish oil have even been reduced to around 600,000 tons. This has led the FAO to consider in their recently-released report on the State of Fisheries & Aquaculture (2012) that "the sustainability of the aquaculture sector will be closely linked with the sustained supply of terrestrial animal and plant proteins, oils, and carbohydrates for aquafeeds". For this reason, Food, Feed and Ceuticals will overcome biofuel applications relating the use of microalgae as a new crop. Due to the fact that some microalgae have a similar composition to fish oils with high levels of omega-3 and omega-6 polyunsaturated Fatty acids (PUFA) like DHA and EPA, they could be used as a replacement for the unsustainable fish oil industry and in mixes with low cost vegetable oils in order to increase their nutritional value.

3. The Current Status and Evolution Pathway in Microalgae Technologies

Table 1.1. Key drivers for the microalgae technologies evolution pathways (2015).

Microalgae market positioning trends

Soy: oil & meal	Fish: oil & meal	Algae: oil & meal
> 200 million ton/year of production	> 7 million ton/year of production	> 20 thousand ton/year of production
Feed applications are the most relevant	**Feed** applications are the most relevant	**Food** applications are the most relevant
Soy based feeds improved with fish meal	Fish meal feeds are improved with algae	Algae emerge as pre-mix feed ingredient
Current value < 0.5 € typically 0.35 €/Kg	Current value < 2 € typically 1.5 €/Kg	Current value > 15 € typically 25 €/kg

Microalgae market opportunities trends

Food & Feed	(Nutra & Cosme) Ceuticals	Agriculture & Biofuels
Biomass formulations	Ingredients and Extracts	Ecological management - soil algae
Formulations: paste, frozen, dried	Extraction: super-critical, green solvents, ...	Processes: inoculation for fertilizers, pyrolysis
Spirulina, Chlorella, Dunaliella	Astaxanthin, Beta-carotene	Future biofuel (dominant route)
Aquaculture paste in Europe since 1997	Plankton and related extracts	High level of social hype

Microalgae development trends

Research	Technologies	Products
First microalgae isolated in 1703 (*Tabellaria*)	> First production reactor in 1957 (MIT roof)	> First product Dihe in Lake Chad
First microalgae in space in 1968 (Soyuz)	Floating devices in the ocean (Omega type)	> *Spirulina* consumption by Aztecs
> 20 species scaled-up for production	Lake based technologies (China, Australia...)	Microalgae biomass for Aquaculture
First microalgae genome sequenced 2007	Raceway based technologies (dominant)	*Spirulina* biomass in a wide range of foods
% based Redfield phytoplankton formula $C_{53.5} H_{7.4} O_{28.2} N_{9.4} P_{1.3}$	Photobioreactors (many configurations)	Microalgae extracts for Ceuticals
Phytoplankton as main CO_2 fixation source	Heterotrophic fermenters (first Celsys in 1985)	Diatoms used in a wide range of industries

Table 1.1 cont. ...

... Table 1.1 cont.

Microalgae academy trends in Europe

Algae Parks	Research Groups	PhD Students
≥ 3 demonstration technology platforms	> 200 research groups (2013)	≥ PhD thesis with 50% microalgae related topics
More than 10 small Parks in Europe (2013)	Average researchers per group ≥ 10 researchers	> 100 PhDs under development (2015)
Reduced networking between the different Parks	Existing but reduced inter-collaboration	Networking limited to Marie Curie projects
Still in a very early stage (reduced training)	Some groups with > 30 years' experience	Diversification of topics applied is increasing

4. Microalgae Products and Markets

The trend is that the 'microalgae sector' will be increasingly market driven—both the actual market and the potential or expectable market (as with biofuels). Product developments will emerge as a result of this market pressure based on existing and developing technology capacity. The current applications will become more relevant as microalgae biomass and microalgae extracts will become possible price-competitive options for formulator's of food, feed, and ceutical products. Biorefinery of microalgae biomass will make possible to have added value products that are competitive in price, quality, and performance.

The products from microalgae have currently only three possible forms (Table 1.2), each of them with a wide range of formulations. Algae paste can be in different concentrations, usually from 15 to 25%, but strongly dependent on the microalgae species and application. Microalgae are usually spray-dried or freeze-dried, but can also be just sun-dried. Extracts can be obtained with solvents—super-critically or just with mechanical processes.

There are two possible ways to evaluate the market size for microalgae products. The more usual one is to determine the approximate production quantities (ton/year) of each microalgae, and consider its average value per kg. An alternative is to consider the top 100 companies, and relate their turnover with the products they have. This is possible and relatively accurate because most companies in the sector are single-product-based, and the top 100 represent more than 80% of the market (Table 1.3). This evaluation was made considering the 2015 annual report, obtained through D&B/Hoovers for the top 100 companies. The value for the global market is calculated accordingly with that balanced size.

Table 1.2. Possibilities of microalgae biomass applications.

Only 3 possibilities:	Paste	Dried	Extracts
	Aquaculture	Food & Feed	Ceuticals
Applications:	–	Aquaculture	–
	–	Ceuticals	–

Table 1.3. Microalgae products global market value.

Microalgae business based on turnover of the top 100 companies—values for 2015

Number of companies	Markets	Segments	< M US$/ year >
	Aquaculture (~ 2%)		
	Paste	*Nannochloropsis* and other	25
	Freeze & spray Dried	*Nannochloropsis* and other	5
		Sub-total	30
	Food & Feed (~ 41%)		
	Dried microalgae	*Spirulina*	300
	Dried microalgae	*Chlorella*	150
	Dried microalgae	Other (including AFA)*	100
		Sub-total	550
	Ceuticals (~ 57%)		
	Extracts: Pigments	Beta-carotene from *Dunaliella*	100
		Astaxanthin from *Haematococcus*	80
		Other	20
	Extracts: PUFA	DHA from *Schizochytrium, Ulkenia and other*	500
		Other (including new *Crypthecodinium*)	20
		Sub-total	720
		Total	1.225
	'Projects'		
	Biofuel related		50
	Other		30
	Microalgae Business	Total Value	1.305

It is a good estimation to consider that Microalgae biomass and extracts have a value of 1.5 million US$/year worldwide.

(1) Government programs to establish new species of microalgae and microalgae extracts, as Novel Foods will be critical to establish them as new crops. The costs for such requests are high, and benefit the complete sector. This regulatory approval is mandatory for most added value applications.

(2) Super-critical extraction and other green/environmental balanced technologies— dissemination and wide use—are essential for the valorization of microalgae extracts for food and ceutical applications.

* *Aphanizomenon Flous Aqua*

5. Microalgae for Biofuels: The Emerging Trends for Sustainable Solutions

From a wide range of possibilities, six technologies are emerging as main drivers for biofuels able to support the economical production of microalgae-based fuels. They comprise microalgae biorefinery for food, feed, fertilizer, and energy production; biofuel production from low-cost microalgae grown in wastewater; biogas upgrading with microalgae production for carbon–neutral, low-cost production of electricity; hydrocarbon *milking* of modified *Botryococcus* microalgae strains; hydrogen production through microalgae biophotolysis; and direct ethanol production from cyanobacteria. They have different possible LCA frameworks and requirements for development and scale-up, and there are many companies and research groups working along these possibilities (Table 1.4).

Table 1.4. Microalgae by-products.

Selected microalgae	Biofuel production from low-cost microalgae grown in wastewater	Biogas upgrading with microalgae for carbon neutral low cost production of electricity	Microalgae biorefinery for food, feed, fertilizer and energy production
GMO microalgae	Direct ethanol production from cyanobacteria	Hydrocarbon *milking* of modified *Botryococcus* microalgae strains	Hydrogen production through microalgae biophotolysis

5.1 Biofuel Production from Low-cost Microalgae Grown in Wastewater

The use of microalgae for treatment of municipal wastewater has been under study for more than 50 years, although it has been used together with physical and chemical removal methods. Microalgae have been shown to be particularly tolerant to many wastewater conditions, removing large amounts of N and P. The necessary CO_2 may be provided from nearby industries at low costs. By simultaneously supplying an effective and cheap growth medium for microalgal biomass production, as well as providing a low-cost sewage treatment, the biomass might then be applied for fuels and fertilizers production.

An example of a research project focused on microalgae production in wastewater for energy production is shown in Box 1.1.

Box 1.1

ALL-GAS PROJECT   (www.all-gas.eu)

Example

The All-gas project seeks to demonstrate the sustainable large-scale production of biofuels based on the low-cost cultivation of microalgae. The complete process chain—from cultivation ponds, biomass separation, extraction of oils and other chemicals to the downstream production of biofuels and their use in vehicle fleets—will take place based on a site of 10 hectares, with the goal of wastewater treatment becoming energy self-sufficient (Fig. 1.2).

Objectives

- Integrate the full production chain of algae to biofuels, using wastewater nutrients and residual biomass energy and CO_2 as main inputs.

- Demonstrate that, using an innovative pond enhancement, the algae culture up to 10 ha can have a yield of biomass production close to 100 t/ha/yr.

- Harvest and transform the algae biomass for the production of biodiesel to feed an engine test bed, and evaluate the extraction of some additional chemicals by innovative methods to obtain value-added chemicals with known markets.

- Implement a new concept for co-digestion of about 5000 m³/d of wastewater, together with algae residue, as well as other extraction by-products, to produce CH_4 and CO_2.

- Upgrade the biomethane for vehicle fuel to power up to 200 cars.

- Obtain the necessary CO_2 to reach the enhanced algal yield from gases of external biomass combustion (i.e., sludge from a wastewater treatment plant located in the area), together with internal biomass combustion (digestate from residual algae and wastewater solids), or other agricultural residues.

- Take advantage of the thermal energy from biomass combustion gases for drying and/or generate electricity to power the system.

5.2 Biogas Upgrading with Microalgae Production for Carbon-neutral, Low-cost Production of Electricity

Biogas (a mixture of methane and CO_2) from landfills or other anaerobic digestion (AD) systems can be flown through microalgae cultures in PBR that consume the CO_2. Biomethane can be burnt to generate electricity or heat, whereas the resulting CO_2 may be reintroduced into microalgal autotrophic cultures. The resulting biomass can be biorefined, and the leftovers re-introduced in the system. The final digestate can be used for valuable fertilizers.

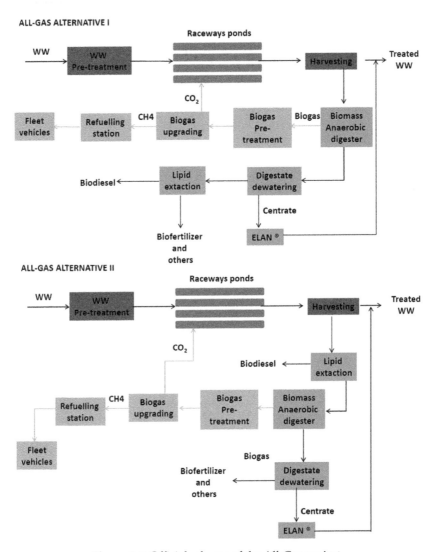

Figure 1.2. Official scheme of the All-Gas project.

5.3 Microalgae Biorefinery for Food, Feed, Fertilizer, and Energy Production

The majority of the mature industrial microalgae producers are focused on a single product application for the biomass, which has a high market value. For products with lower market value, such as biofuels, protein, or animal feed, the economic viability of the process is only achieved within the logic of biorefinery, meaning that all its secondary products or biomass residues are recovered and applied on the market. Economic viability can be achieved by having biofuel application as a secondary objective.

An example of a research project focused on microalgae production through a biorefinery concept for energy production is shown in Box 1.2.

Box 1.2

BIOFAT   (www.biofat-project.eu)

Example

BIOFAT was a 'microalgae to biofuel' demonstration project aiming to integrate all the processes from single cell to biofuel production, and show what requirements and performance are necessary for a yearly productivity in the order of 100 ton/year at a 10 ha scale. In order to achieve those results, two pre-industrial scale Pilot Plants of 0.5 ha each were built and operated: one in Italy, and one in Portugal. Based on the accumulated knowledge and experience of the consortium partner companies and universities, the BIOFAT developed and implemented one of the first sustainable industrial scale-up strategies for microalgae biorefinery worldwide. This resulted in an attractive business case for a 10 ha DEMO facility. Encompassing the entire value chain, BIOFAT partners have worked out the main bottlenecks that were still hindering the full industrial application of such a promising field: by developing novel cultivation and harvesting technological solutions, the BIOFAT consortium was able to develop a business case that is economically sound, environmentally sustainable, as well as technologically robust.

The two pilot plants that operated during the project (pre-industrial scale) showed that it is possible to compare the results of microalgae biomass productivity, composition for biofuels, and other added value products (namely the omega-3 fatty acid EPA and other products rich in protein and fatty acids, with application in the food and feed markets). Using a proxy model developed during the project, it is possible to simulate the performance for any location. With this proxy, which correlates the solar radiation with the microalgae productivity, it was possible to extrapolate the annual microalgae biomass productivity of a DEMO plant with data from the pre-industrial Pilot Plants. It was also possible to provide a comprehensive, robust, reproducible, and industrially appealing set of experimental results for an attractive Business Case development. The final 10-ha DEMO Plant was designed so that the same production obtained in the Pilot Plants can be guaranteed, and the scale-up risk is low, consisting in 'a set of modules equal to the pilot plants'.

Several LCA studies were performed during the project. The LCA results from both Pilot Plants were presented, as well as those for the several scenarios considered for the DEMO Plant. Given the obtained results, it was proven from the consolidated operation data obtained in April/2016, that when the microalgae production plant includes a CHP (Heat and Power generation via combustion) system, fed with a non-fossil source, the algae biofuels production is a net carbon fixation industry. Moreover, in the framework of the Algae Cluster

---

Box 1.2 cont. ...

... Box 1.2 cont.

(joint platform of the 3 FP7 funded projects: BIOFAT, Intesusal and All-gas) the LCA study was peer reviewed and validated by an external LCA specialist. Work is already underway to compare, under an Algae Cluster mutually agreed methodology, not only the impact of BIOFAT in terms of GHG emissions, but also other parameters. The result of this joint analysis will be published after the end of the 3 projects.

The BIOFAT Business Case for a 10-ha DEMO Plant resulted in the following case scenarios (all scenarios evaluated in a 20 year period): (1) bioethanol production from *Tetraselmis*, with biomass yield of 30.8 ton/ha/year and bioethanol yield of 3.8 ton/ha/year, with breakeven achieved after 12 years, 7.4 M€ NPV (net present value) and IRR (Internal return rate) of 14%; (2) *Nannochloropsis* production in Nitrogen (N)-depleted cultivation, where the main product is biodiesel and the remaining biomass is applied for added-value products—this scenario has a yield of 25.5 ton/ha/year of biomass and 12.7 ton/ha/year of biodiesel, but both IRR and NVP are still negative after 20 years; a sensitivity analysis of this scenario indicates that for biomass productivity above 75 ton/ha/year (locations with a higher radiation profile than South Europe), equivalent to a production of biodiesel of 37.4 ton/ha/year, the IRR increases to 13%; (3) *Nannochloropsis* production, but with N-replete cultivation, where the main products are added-value compounds, and biodiesel is a secondary product—in this scenario the biomass yield is 37.8 ton/ha/year and the biodiesel yield is 3.4 ton/ha/year; it has achieved breakeven point after 8 years of project, with 16.4 M€ NPV and an IRR of 21%; (4) a scenario for only added-value products from both strains (for the markets of feed and food), with a yield of 25.5 ton/ha/year of *Nannochloropsis* and a *Tetraselmis* yield of 35.2 ton/ha/year, with quite attractive financial results: discounted payback period between 5 and 6 years and IRR values between 28 and 36%.

The results of the Project are being evaluated by investors and a possible location was already identified for the establishment of the 10 ha DEMO facility which is expected to start construction in 2017, using the knowledge gained in BIOFAT project.

The following tables (Tables 1.5 and 1.6) present the main technologies comprising the 5 microalgae prototype and pre-industrial scale plants that were operated during the BIOFAT project development.

5.4 Direct Ethanol Production from Autotrophic Cyanobacteria

Expressive advances in the development of genetic tools to increase energy production in microalgae and cyanobacteria have recently been achieved, and are being used to manipulate central carbon metabolism in these organisms. Manipulation of metabolic pathways can redirect cellular functions towards synthesis of preferred products. It is likely that many of these advances can be extended to industrially relevant organisms. Several companies and research groups have already established the proof of

Table 1.5. Five prototype plants of different production area were operated during the project.

A4F Portugal, Lisbon	A4F Portugal, Pataias	F&M/UNIFI Italy Florence	A&A Italy, Camporosso	BGU Israel, Beersheba
Prototype plant of A4F located in Lisbon. Gave support to the BIOFAT Pataias Pilot Plant operation through R&D activities	Prototype plant in Pataias. The generated data was used to design the BIOFAT Pataias Pilot Plant and improve its operation	Prototype plant in Florence. The generated data was used to design the BIOFAT Pataias Pilot Plant and improve its operation	Prototype plant in Camporosso. The generated data was used to design the BIOFAT Pataias Pilot Plant and improve its operation	Prototype plant in Israel. Providing troubleshooting for all partners and future insights
Green Wall Panel (1000 L)	Green Wall Panel (1000 L)	F&M bubble columns	Green Wall Panel (GWP®-II) 2 x 8000 L, 2 x 250 m	Green Wall Panel reactors (400 L, 8 m, 50 L m^{-1})
Tubular Photobioreactors (4 x 1000 L and 2 x 1500 L)	Tubular Photobioreactors (3 x 6000 L)	Green Wall Panel (GWP®-II) and Open Ponds	Improved Raceway ponds (2 x 50000 L, 530 m^2 2 x 120000 L, 1280 m^2)	Small Raceway Ponds (2,5 m^2)

Table 1.5 cont. ...

... Table 1.5 cont.

A4F Portugal, Lisbon	A4F Portugal, Pataias	F&M/UNIFI Italy Florence	A&A Italy, Camporosso	BGU Israel, Beersheba
	Cascade Raceway (15 m², 900 L)			Raceway ponds (5000 L, 40 m²)

Table 1.6. Pre-Industrial Scale Plants.

Portugal, Pataias—BIOFAT Pataias Pilot Plant (BPPP)	Italy, Camporosso—BIOFAT Camporosso Pilot Plant (BCPP)
Cultivated strains: *Nannochloropsis oceanica* F&M-M24	Cultivated strains: *Nannochloropsis oceanica* F&M-M24; *Tetraselmis suecica* F&M-M33
Inoculum production in GWP-I at BPPP (3000 L)	Inoculum production in GWP®-II at BCPP (2 x 16000 L)

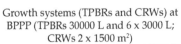

Growth systems (TPBRs and CRWs) at BPPP (TPBRs 30000 L and 6 x 3000 L; CRWs 2 x 1500 m²)	Growth systems (GWP®-II modules and RWs) at BCPP (2 x 500 m², 2 x 1300 m²)

Table 1.6 cont. ...

... Table 1.6 cont.

Portugal, Pataias—BIOFAT Pataias Pilot Plant (BPPP)	Italy, Camporosso—BIOFAT Camporosso Pilot Plant (BCPP)

Pre-concentration ceramic (left) and capillary (right) membrane filtration system	Pre-concentration membrane filtration system

concept, and the scale or inclusion of co-products looks to be the requirement for a positive economic balance.

An example of a research project focused on microalgae production for direct ethanol production is shown in Box 1.3.

Image 1.3 presents the main technologies comprising the prototype plant operated during the DEMA project development, which is a GMO compliant facility.

5.5 Hydrocarbon Milking of Modified Botryococcus Microalgae Strains

Among the costly downstream processing steps, it is generally accepted that harvesting/dewatering, and the following extraction of fuel precursors from the biomass are costly and energy-demanding steps. In order to decrease their cost, the possibility of integrating the steps of harvesting and extraction seems to be a potential solution. In this integrated process, algal cells can be cultured in biphasic bioreactors, consisting of an aqueous phase, where algae grow, and an organic phase, immiscible with the former, which continuously extracts the fuel precursors. Genetically modified *Botryococcus*, which naturally excretes biofuel precursors is being studied for commercial application in Japan and the US.

5.6 Hydrogen Production Combining Direct and Indirect Microalgae Biophotolysis

Photobiological hydrogen production has advanced significantly in recent years, and nowadays a variety of photosynthetic and non-photosynthetic microorganisms, including unicellular green algae, cyanobacteria, anoxygenic photosynthetic bacteria, obligate anaerobic, and nitrogen-fixing bacteria are endowed with genes and proteins for H_2-production. These organisms can use two natural pathways for H_2 production: (i) H_2-production as a by-product during nitrogen fixation by nitrogenases; and (ii) H_2-production directly.

Box 1.3

DEMA   (www.dema-etoh.eu)

Example

The Direct Ethanol from Microalgae (DEMA) project, financed by the EU, congregates 9 European partners: University of Limerick (Ireland); Universiteit Van Amsterdam, Photanol BV and Pervatech BV (Netherlands); University of Cambridge and Imperial College of London (United Kingdom); Ercane GIE (France) and LNEG, A4F-Algae for Future (Portugal) (Fig. 1.3), leaders in their respective research fields, to developed a novel strategy based on a series of complementary innovative methodologies to autotrophically cultivate a cyanobacterial strain—*Synechocystis* sp.—that is capable of directly producing ethanol at an economically viable energy balance. Indeed, the main goal is to develop, demonstrate, and license a complete economically competitive technology for the direct production of bioethanol from microalgae with low-cost scalable photobioreactors by 2016. For the DEMA project microalgae cells are seen as catalysts, i.e., single cell factories producing and excreting ethanol to the medium for later extraction and purification. Several academic research facilities and one start-up enterprise have developed initial proof-of-concepts in this domain, but the technology still faces major challenges before large-scale commercial production can be realized.

A4F is cultivating engineered ethanol producing strains in a unique GMO compliant Experimental Unit located in Lisbon (UEL), with closed photobioreactors, optimized sun exposure, in liquid medium supplemented with CO_2, nitrogen, phosphate, and micronutrients to offer the means for direct synthesis of bioethanol from sunlight. The Lisbon Experimental Unit also allows for media recycling and optimization, which is a critical parameter for industrial deployment. Regarding the production system, a pilot scale PBR was developed and constructed for the DEMA project, taking into account the materials used and the coupling to ethanol extraction and purification unit. At A4F, it is possible to cultivate and optimize the microalgae production at laboratory scale and pilot scale—currently up to 1000 L. The ethanol quantification is carried out by gas chromatography at DEMA partner LNEG.

Ethanol production was achieved, currently at concentrations of ethanol in the culture media of 1 g/L at laboratory scale and 0.4 g/L at pilot scale in a controlled non-axenic environment. Additional work is being carried out by the DEMA consortium to improve the technology, the strain capacity to produce ethanol, and focusing in optimization of the culturing conditions, namely, pH, culture media, and temperature.

Experimental Laboratorial Unit (UEL), Lisbon, Portugal

Tubular Multilayer PBR (vertically stacked tubes) - 1100 L, 15 m²   Flat Panel PBR (4 cm optic path) - 55 L; 1 m²   Tubular Unilayer PBR (horizontally distributed) - 200 L; 3.5 m²

Figure 1.3. Images of the Experimental Unit (UEL) photobioreactors used in the DEMA project, in Portugal, Lisbon.

6. Microalgae Species

The trend is that the microalgae that generate more publication of digital documents are the ones that are recognized for human consumption: *Spirulina* and *Chlorella*; surprisingly new applications brought microalgae as *Navicula* and *Euglena* to capture a relevant interest; and *Chlamydomonas*, being a model organism is also a very relevant microalgae, where relevant knowledge has been accumulated.

A possible way to measure the existing 'body of knowledge' and its evolution is through the 'Google Index', which is the number of references existing in a Google search in a specific moment (can increase or decrease according with web movements, such as the database open for searching engines). This is relevant if developed in a comparative way, and if the numbers of references are large so that the issues related with the quality of information can be neutralized. This represents the evolution of interest in the different genera of microalgae, both at the research and production levels.

In the top 30 list, it is possible to see that the following microalgae had more than 100% growth in the last 10 years: *Thalassiosira, Navicula, Gymnodium, Emiliana, Pavlova, Haematococcus, Nannochloropsis, Botryococcus, Phaeodactylum*. Evaluating the remaining of the top 70 list, the following microalgae also show a growth of more than 100%: *Dysmorphococcus, Haslea, Neochloris, Schizochytrium, Chlorococcus* (not presented here). The 'long tail' also has several interesting future tags. It is very relevant to see that in 2013, the top 5 species represent more than 80% of the references.

If the newcomers into the sector want to get knowledge and experience from marine bivalve hatcheries where microalgae are produced in a m³ scale, they will find the following microalgae that represent 90% of the manager choices: *Isochrysis, Chaetoceros, Tetraselmis, Pavlova, Skeletonema.* Over the last four decades, several hundred microalgae species have been tested as food, but probably less than twenty have gained widespread use in aquaculture.

This is a top 25 list based on an analysis from the universe of 70 genera analyzed (Table 1.7). The Google search 10 years ago was done in the month of September, as well as in 2013. The objective was to evaluate the comparative relevance, but now it is possible to study the evolution of the interest level (growth in the number of references); the current relevance level, and the evolution of relevance level. Searches combined as simultaneous keywords the name of the genera and algae or cyanobacteria. This kind of methodology has been used in other fields with interesting results, as it provides information about the number of documents with the specified references.

7. Microalgae Production Options

The trend is that depending on the microalgae crop and the production platform, there will be different production options or carbon sources. Most microalgae production is now photoautotrophic (*Arthrospira, Chlorella, Dunaliella...*), representing 80% of production in ton/year or heterotrophic (*Schizochytrium, Ulkenia, Cryphecodinium...*), representing 20% of production in ton/year. Mixotrophic production will become a third option, and in the near future it is expected that each of the referred production options will have an equivalent importance in terms of production.

The microalgae nutrition mode has a high impact in the productivity, in production costs, and in the possible production platforms that can be used (Table 1.8).

The term 'micro-algae' is usually used in its narrowest sense as a synonym for photoautotrophic, unicellular algae utilizing CO_2, and gaining energy from light through photosynthesis. Although certain species are obligate photoautotrophs, numerous microorganisms are capable of both heterotrophic and photoautotrophic metabolism, either sequentially or simultaneously. Some even prefer to live heterotrophically, and only survive as photoautotrophs once all other energy sources are depleted. These microorganisms may thus not be classified as pure microalgae. There is a relevant debate in the scientific community about this, which may be resolved with the increase of genetic information available on these species.

7.1 Autotrophs can be Phototrophs or Chemotrophs

Autotrophic species are photosynthetic like plants. Phototrophs use light as an energy source, while chemotrophs utilize electron donors as a source

Table 1.7. Ranking evolution along 10 years in scale Google of references for microalgae genera.

Genera + "alga"	Class	Google 2003	Google 2013	fold-change in 10 Y	Scholar 2013
Spirulina (*Arthrospira*)	Cyanophyceae	303,000	8,833,000	29	57,220
Chlorella	Chlorophyceae	104,000	4,780,000	46	134,000
Navicula	Bacillariophyceae	271	1,290,000	4,760	31,400
Chlamydomonas	Chlorophyceae	28,000	1,170,000	42	52,900
Euglena	Euglenophyceae	18,800	1,020,000	54	48,600
Anabaena	Cyanophyceae	22,100	789,000	36	61,800
Synechocystis	Cyanophyceae	117,000	652,000	6	38,200
Dunaliella	Chlorophyceae	8,330	634,000	75	32,200
Nostoc	Cyanophyceae	20,400	593,000	29	36,100
Scenedesmus	Chlorophyceae	11,800	419,000	36	79,100
Aphanizomenon	Cyanophyceae	6,660	333,000	50	12,500
Haematococcus	Chlorophyceae	1,800	327,000	182	11,200
Oscillatoria	Cyanophyceae	6,520	301,000	46	30,600
Nitzschia	Bacillariophyceae	8,300	298,000	36	33,800
Thalassiosira	Bacillariophyceae	47	273,000	5,809	18,600
Skeletonema	Bacillariophyceae	5,140	211,000	41	21,300
Chaetoceros	Bacillariophyceae	5,590	209,000	37	24,500
Phaeodactylum	Bacillariophyceae	1,730	192,000	111	12,000
Selenastrum	Chlorophyceae	5,660	192,000	34	12,400
Emiliana huxleyi	Haptophyceae	310	186,000	600	11,900
Nannochloropsis	Eustigmatophyceae	1,080	180,000	167	9,930
Gymnodium	Cyanophyceae	166	173,000	1,042	547
Cyclotella	Bacillariophyceae	4,640	157,000	34	25,600
Pavlova lutheri	Cyanophyceae	277	154,000	556	4,830
Botryococcus	Chlorophyceae	1,150	143,000	124	9,790
Tetraselmis	Prasinophyceae	1,880	127,000	68	12,200

of energy, whether from organic or inorganic sources. However, in the case of autotrophs, these electron donors come from inorganic chemical sources. Such chemotrophs are lithotrophs. Phototrophic prokaryotes may utilize a variety of carbon sources, depending on the metabolic pathways available.

Table 1.8. Possible production platforms for microalgae production.

only 3 possibilities:	Photoautotrophic	Heterotrophic	Mixotrophic
production platforms:	Ocean Lakes Ponds PBRs Laboratory	Fermentors Laboratory	Ponds PBRs Laboratory
	CO_2 as the carbon source	Organic Carbon as the Carbon source	CO_2 and Organic Carbon as Carbon sources

7.2 Heterotrophic

Heterotrophic species get their energy from organic carbon compounds, in much the same way as yeast, bacteria and animals. Cultivation without light and with the controlled addition of an organic source of carbon and energy is similar to procedures established with bacteria or yeasts in multipurpose stirred closed tanks sterilized by heat. To date, only a small number of microalgal species have been cultured heterotrophically in conventional bio-reactors. Typically, heterotrophic cultures can have 5 to 10 times higher densities than corresponding autotrophic.

7.3 Mixotrophic

Organisms deriving nourishment simultaneously from both autotrophic (inorganic substances resulting from chemosynthesis and photosynthesis) and heterotrophic (organic substances) mechanisms. Chlorophyll-bearing flagellates, which are autotrophs, become mixotrophic in heavily polluted water, where they feed on organic matter in order to stimulate growth and reproduction; some of these flagellates can develop even in total darkness, that is, without photosynthesis. Mixotrophic species can use both sunlight and organic carbon, whatever they can get.

Mixotrophic cultures started to be used in Japan and Taiwan in mid-60s, but for the past many years, the autotrophic production has been preferred because of lower production costs and easier production operation due to contaminants. Only recently, the need for higher productivities with added value microalgae re-started this approach for large-scale production of microalgae.

The following Box 1.4 presents the example of a biotechnology company based in Portugal with activity in the microalgae field, where several of the technologies described are designed and used for microalgae production and R&D activities.

Box 1.4

A4F—ALGAE FOR FUTURE   (www.a4f.pt)

Example

A4F—Algae for Future incorporates the knowledge and experience of more than 20 years in the field of microalgae biotechnologies, including research projects. It is a bioengineering company which designs, builds, operates, and transfers microalgae biomass industrial production plants worldwide. In order to maximize the performance of each process, A4F proposes an approach through a gradual scale-up from laboratory inocula up to the large-scale final plant. A4F has a unique track record of industrial production of more than 15 different strains of microalgae, ranging from freshwater to saline or hypersaline environments, using autotrophic, mixotrophic, or heterotrophic growth regimes, and applying the most appropriate, and industrially proven types of systems, such as green wall panels, tubular photobioreactors, conventional raceways and cascade raceways, or fermenters (Fig. 1.4).

A4F owns an R&D laboratory that supports experimental pilot unit located in Lisbon (UEL), where trials are performed at pilot scale. The laboratory is fully equipped and licensed for the maintenance and scale-up of genetically modified microorganisms (GMO's), licensed by APA (Official Portuguese Agency for the Environment) for class I and II GMO handling and cultivation. The R&D laboratory is equipped with all the necessary material to develop an experimental work, including growth chambers with controlled temperature and light, general laboratory equipment, and molecular biology equipment (PCR, optical and fluorescence microscopy, etc.). In order to comply with existing regulations, the laboratory has a Laminar Flow Chamber class II and different equipment specific to GMOs. The UEL unit is modular, GMO compliant and has currently 1000 m^2 of capacity, including 70 m^2 photobioreactors (PBRs) and the total capacity will soon be expanded up to 2000 m^2. Geared towards flexibility, this unit has the capacity to produce (from cell to final product (powder or other)) virtually every species of microalgae/cyanobacteria, using autotrophic or mixotrophic growth conditions, using fresh or saltwater, and also to process the culture into a concentrate or a dry powder.

In the GMO compliant pilot unit, each PBR can work separately with a dedicated water supply allowing cultivating fresh and salt water species simultaneously. The unit has 9 operating systems, each one responsible for a resource or process within the plant's process flow. This unique process design successfully integrates feedstock supply, algae cultivation, production and biomass processing. The operating systems are: water treatment and distribution, nutritive medium preparation and distribution, carbon supply, temperature control, harvesting, biomass processing (spray-drier), effluents treatment and control system and automation.

Furthermore, A4F's facilities are located within the campus of a national R&D institute for energy and geology shared with many technological SMEs. As a result of a contract established between A4F and the institute, A4F has access to complex analytical techniques such as chromatography (HPLC and GC), flow cytometry with cell sorting, fermentation equipment at pilot scale, supercritical extractors, among others.

Box 1.4 cont. ...

... Box 1.4 cont.

> A4F combines these capabilities with relevant international market knowledge and experience on microalgae marketing and sales, as well as product development. Furthermore, A4F cooperate with our clients and partners in researching and developing new, novel and innovative microalgae based products and applications. Also, A4F supports the design and implementation of marketing strategies for economic exploitation and commercialization of microalgae products and applications in the global market. Our concept is the development and production of different species of microalgae, and microalgae-based products through biorefining, on an industrial scale and in a sustainable and commercially profitable way, through different technological solutions, from different needs, with different partners and for different uses: food, feed, fiber, fertilizer, pharma, and fuel, as a contribution to a sustainable future.

8. Microalgae Production Platforms

The trend is that different microalgae species will be produced in different platforms in different locations. As with other crops such as soy, sunflower or corn, not all microalgae can grow in any climate. All six possible production platforms will have a different applications framework. During the next decade will be possible to see a consolidation of open ponds and PBRs and a massification in fermentation-based companies. Ocean-based production will emerge as a possibility for special places, and Laboratory-based 'molecular farming' will be common.

Two types of production systems will distinguish the microalgae crops from one another: open system based platforms will be agriculture related and will follow agriculture approaches, while closed system based platforms will be industry related. Different species will be produced according with the objectives and available resources (Table 1.9).

Production platforms are of utmost importance in microalgae production, because their shape and operational conditions will dictate the final productivity rates achieved. The choice of the most suitable platform is case dependent, as well as the production mode.

8.1 Ocean

This possible production platform is still in a research stage. Several research groups and at least 3 companies have been working in possible approaches to grow microalgae in floating devices, and built proof of concept systems. First experiments started along the last ten years and the most well known is the Offshore Membrane Enclosure for Growing Algae (OMEGA), aiming to demonstrate, on one hand, the viability and scalability of producing large amounts of algae for carbon-neutral biofuels, foods, fertilizers, and other

Lisbon Experimental Unit (UEL), Portugal, Lisboa

1000 m² GMO compliant pilot unit

Analytical and inoculum production laboratory

Water treatment, culture medium preparation and distribution and carbonation systems

Cultivation capacity of 10 m³ with open systems (Flat panel, tubular fence photobioreactor, unilayer photobioreactor) and open systems (Cascade Raceways).

Processing facilities: centrifugation, spray-drying

Culture thermoregulation

Water recirculation, wastewater treatment and other utilities

Monitoring and control systems

Figure 1.4. Images from the A4F—Algae for Future, Lisbon Experimental Unit (UEL), Lisboa, Portugal.

Table 1.9. The different types of microalgae production systems.

Open:	Ocean	Lakes	Ponds
Closed:	PBRs	Fermentors	Labscale

valuable products, and on the other hand, treat wastewaters and sequester CO_2, without competing with traditional farming for land, fresh water, or fertilizers. Large amounts of algae-producing units, composed of floating bags in the sea, would sequester CO_2 and use seawater as growth medium (http://inhabitat.com/nasas-omega-project-creates-carbon-neutral-food-and-fuel).

8.2 Lake

This was the first approach for the human use of natural microalgae blooms for food consumption. Harvesting of naturally occurring microalgae was reported more than 500 years ago both in Mexico and in Lake Chad. *Spirulina* blooms are still used for commercial application in China and Myanmar—and *Dunaliella* in Australia. The technology to cultivate *Spirulina* using natural lakes was first developed in the early 1970s in Lake Texcoco in Mexico (stopped operation in the 80s), and for *Dunaliella* it started in the 80s and is still in use in two locations in Australia for the production of beta-carotene and other carotenoids.

8.3 Ponds (Circular, Raceways and Other)

This is the most widely used production platform with raceway ponds, circular ponds, or other configurations. Chlorella Industry Co., Ltd., headquartered in Tokyo, pioneered the world's first mass culture production in 1964. These facilities used circular ponds to grow *Chlorella* for human consumption. In 1966, Taiwan Chlorella Company was designed and built by Japanese experts. In the early 70s, *Spirulina* production using raceways started in the hot desert area in the southeastern part of California (Earthrise) with a commitment to developing microalgae for food, biochemicals, and pharmaceuticals. These companies were role models for many others and knowledge spread all over the world, mostly in China for the production of *Spirulina* and *Chlorella*.

8.4 PBRs (Many Possible Configurations but Tubular Systems are the Most Common)

The first successful industrial tubular PBRs were established by Algatechnologies in Israel in 1999, to develop and commercialize astaxanthin for the Ceutical industry. IGV Germany also started relevant work with different types of PBR at this time, and through the years, a wide range of systems have been proposed.

8.5 Fermenters

Martek Biosciences Corporation (now acquired by DSM) developed and patented several fermentable strains of microalgae, which produce oils rich in docosahexaenoic acid (DHA) in fermenters that range in size from 80,000 to 260,000 liters. It is at the present the largest value business related with microalgae.

8.6 Labscale

This platform and scale will become relevant for specialty, high value products in the concept of 'molecular farming'. Labeled isotope extracts are already a small business at this level. GMO microalgae will be used in most of the cases. Some of these molecules such as the diterpenoid targets in this project have market values of well over €10 million/kg.

9. Microalgae Production Scale

The trend is that depending on the microalgae crop/market need driven, and the corresponding production platforms, there will be a different choice of production options—and production scale. The evolution is happening in all different scales, but there is a strong emergent MicroFarming movement, starting in France—and a Large-Scale Farming in US and China. High potential in the 3As—Australia, Africa, and Arab countries, may lead to new ventures in these countries in the next 10 years. Production scale, novel food approval, and marketing about microalgae species will shape the trend at this level (Table 1.10).

Culturing microalgae in the laboratory is 150 years old, and commercial farming just 60 years old. Archaeological data indicates that the domestication of various types of plants and animals evolved in

Table 1.10. The several microalgae production scales.

	Large-scale farming	Regular Farming	Microfarming
Open systems:	> 100 ha	> 10 ha	> 1 ha
Closed systems (PBRs)	> 1.000 m^3	> 100 m^3	> 10 m^3
Industrial biotechnology	Large-scale fermentation	Regular fermentation	Lab & Pilot fermentation
Closed systems (Fermentors)	> 1.000 m^3	> 100 m^3	> 10 m^3

separate locations worldwide, starting around 12,000 years ago. Neolithic Revolution, sometimes called the Agricultural Revolution, was the world's first historically verifiable revolution in agriculture. Microalgae will be the 21st century's new agriculture crop. Micro-crops with hydroponics and related technologies will become standard technologies all over the world.

There is no optimal unique solution for microalgal production. However, several approaches have been developed by researchers, either by using alternative approaches, improving technologies in order to bypass the costly harvesting and extraction steps, and operating in a general overview concept of biorefinery, as happens for example with soy, where soy meal and soy oil are combined for global price sustainability. Integrated systems, with several technological options combined in the same operation unit, may provide increases in productivity and reduction in costs; examples are the utilization of closed photobioreactors for the production of higher quality inoculation (5–10% of total production volume) and 90 to 95% in open systems as raceways, ponds, or cascade systems. These system configurations must always be adapted to the microalgae species cultivated, final market features and production unit location.

The scale of production is still related with price performance and consideration as crop (Table 1.11).

The marketing is also a relevant bottleneck for commercialization as would be for other crops if the scientific names were used in the trade (Table 1.12).

10. Microalgae Research Needs

The trend is that evaluating the 'gaps' and problems in the scale-up of microalgae technologies, the need is increasingly flowing for 'industrial ecology' related issues in what concerns biology related trends—this is mostly because contamination outside the laboratory is a key limitation. Technology related needs will mostly be in the discovery that it is critical to adapt and use processes that are already routine in other industries. At

Table 1.11. Tons of protein/lipids produced by year.

Protein/Lipid	Soy oil & Soy meal	Fish oil & Fish meal	Algal oil & Algal meal
Production Scale (ton/year) 2011	200.000.000s	7.000.000s	15.000s

Table 1.12. Different names used in the commercialization.

Cattle food	*Medicago sativa, Trifolium, Poaceae …*	alfalfa, red clover, graminae...
Fruits	*Malus domestica, Pyrus, Prunus persica, Prunus, Citrus sinensis …*	apple, pear, peach, orange,...
Grains	*Triticum, Zea mays, Secale cereale, Avena sativa …*	wheat, corn, rye, oat,...
Microalgae	**Arthrospira, Chlorella, Haematococcus, Dunaliella, …**	*Spirulina,...*
Nuts	*Cocos nucifera, Arachis hypogaea, Carya illinoinensis …*	coco, peanut, pecan
Other crops	*Gossypium, Rosa rubiginosa …*	cotton, sweet briar,
Vegetables	*Solanum lycopersicum, Lactuca sativa, Spinacia oleracea …*	tomato, lettuce, spinach

the same time *analytical* processes for fast and accurate information are essential to manage the production of microscopic organisms in a large scale which is in a level that is 1,000 times smaller than humans were used to work, and where there are important need when comparing with other agriculture crops.

Research needs are a 'best practice' item in review papers or other publications where results were almost able to show the expected hypothesis. Under the Project AQUAFUELS, Vítor Verdelho coordinated a team with Rene Wijfels that evaluated several hundreds of 'research needs' statements through publications in the last 20 years [AquaFuel FP7 Coordination Action. FP7 ENERGY—2009-1. Algae and Aquatic biomass for a sustainable production of 2nd generation biofuels. Task 3.1. Research Needs. January 2011. See www.aquafuels.eu for details]. This meta-analysis work with information from direct experience in microalgae biotechnologies, both academic research and industrial scale-up, led to the following synthesis about applied research needs (Table 1.13).

It is possible to consider that Applied Research Needs should be connected with Market Research needs and trends. However scientific research is not connected with specific needs, but rather with financing opportunities and specific academic interest related with career development (Table 1.14).

Table 1.13. Bio related and Technology related aspects.

Bio related			Tech related		
Molecular	Cell	Ecosystem	Cultivation	Processing	Analytical

Table 1.14. Several levels of Applied Research Needs.

Levels	Research needs
Molecular	Annotated genome of the top 10 microalgae genera; complete elucidation of the metabolic pathways for pigments and lipids produced by each algae will have its own lipid profile, thereby it is crucial to utilize species that have a suitable lipid profile for biodiesel production; understand the metabolic factors and regulation mechanisms involved in pigment and fatty acids biosynthesis; metabolic pathways elucidation.
Cell	Cell wall and cell membrane behavior; understanding cell cycles to improve growth; optimizing of nutrition and efficiency of utilization of nutrients; microalgae physiological changes under stress response; information that will reveal the regulation networks that are responsible for microalgae behavior during stress periods.
Ecosystem	Cell-cell interactions: symbiosis w/bacteria (vitamins, hormones...); quorum sensing in microalgae ecosystems; cell death phenomena; biofilm formation and extracellular metabolite production; allelopathy control & contamination management; microalgae—bacteria interactions; the influence of the bacteria in microalgae cultivation must be previously evaluated since they can have a positive or a negative interaction, the presence of bacteria might be beneficial to some algae species by increasing the growth rate due to the establishment of symbiotic relationships; viral infection management will be as important for microalgae cultures as for any other plant crop.
Cultivation	Microalgae cultivation using systems with lower contamination levels, easier to operate and to scale-up; species to be cultivated have to be selected according to stability of the strain; time of generation and culture synchronization; nutritional requirements, yield in oil and fatty acids profile; optimized and specific culture media using stoichiometric/mass balance approaches; nutrients: *chemical species* form interactions and change bioavailability; use of zeolites and other culture media enhancers; water recirculation and dilution strategies.
Processing	Harvesting technologies with membranes for improved recirculation of culture media; Clean processing with specific technologies for algae— including mechanical (pressure...), chemical, biochemical, electrical pulsed electric fields (PEF) and thermal will be extremely relevant; supercritical extraction technologies for pigments and oils; low cost preparative chromatography for purification of specific compounds.
Analytical	Remote sensing with spectroscopy based camera devices will be critical for scale-up; new analytical methods for physical and chemical parameters that are species-specific identification for rapid and specific detection of microalgae; DNA chip for fast identification/screening.

Microalgae production must be considered as a "new crop" which still requires a large effort for scale-up and for "value-chain" management and exploitation. Energy related applications are part of a global opportunity strongly dependent on the production framework.

Keywords: Microalgae; biorefinery; biofuels; ceuticals; aquaculture; photobioreactors

Reference

Chisti, Y. 2007. Biodiesel from microalgae. Biotechnology Advances 25: 294–306.

CHAPTER 2

Use of Wastewater to Improve the Economic Feasibility of Microalgae-Based Biofuels

Ana L. Gonçalves,[1] *Sérgio L. Pereira,*[1,2]
Vítor J.P. Vilar[2] *and José C.M. Pires*[1,*]

1. Introduction

The increase in world population led to the increase of energy consumption to levels that can compromise the economic growth. Current energy sources are highly dependent on fossil fuels. The intermittence of its supply (due to the political instability of the supplier countries) promotes the variability of the fossil fuel price, affecting the energy security of developed countries. Moreover, the intense use of fossil fuels is related to several environmental concerns, such as air pollution and climate change (Gaffney and Marley 2009, Haines et al. 2006, Monks et al. 2009). Regarding climate change, this phenomenon is associated with greenhouse gases (GHGs, mainly carbon dioxide—CO_2) emission. In the last United Nations Climate Change Conference in Paris (COP21), several countries agreed to reduce their GHG emissions to limit the rise in global temperature to less than 2°C (Hulme 2016, Lewis 2016). Main commitments and actions are focused

[1] Laboratório de Engenharia de Processos, Ambiente, Biotecnologia e Energia - LEPABE, Departamento de Engenharia Química, Faculdade de Engenharia, Universidade do Porto, Rua Dr. Roberto Frias, 4200-465 Porto, Portugal.
[2] Laboratory of Separation and Reaction Engineering – Laboratory of Catalysis and Materials (LSRE-LCM), Departamento de Engenharia Química, Faculdade de Engenharia, Universidade do Porto, Rua Dr. Roberto Frias, 4200-465 Porto, Portugal.
* Corresponding author: jcpires@fe.up.pt

towards energy efficiency, renewable energy deployment, and forest protection (increasing the natural sinks of CO_2). In this context, biofuels have the potential to replace fossil-based fuels, being a carbon-neutral fuel. Biofuel production should be performed with non-edible feedstocks to avoid the competition with human food market (Chhetri et al. 2008, Gui et al. 2008).

Microalgal cultivation has recently attracted much attention due to its advantages, such as rapid growth rate, high CO_2 fixation rates, tolerance to a wide range of environmental factors, and the potential as a feedstock for several industries (Chisti 2007, Grima et al. 2003, Hu et al. 2008, Pires et al. 2012, Pulz and Gross 2004, Spolaore et al. 2006). Microalgae are considered as a promising alternative for biofuel production. These microorganisms can grow in places that are unsuitable for agriculture, not competing with land for food production. However, mass culture of microalgae at a commercial scale is only performed for high-value products, due to high production costs. The main factors contributing to the high production costs of microalgal culture include the increase of nutrients' price, which doubled in the last years, and the necessity for freshwater, which also constitutes an environmental drawback (Menetrez 2012, Vasudevan et al. 2012). To have an economically viable process of biofuels production, cost reduction of production and downstream steps is mandatory. Consequently, several researchers have studied the use of nutrient-rich wastewater (in nitrogen and phosphorus) as culture medium, significantly reducing the need for nutrients and freshwater and, at the same time, promoting the treatment of these effluents (Pires et al. 2013, Rawat et al. 2011). In this context, this chapter aims to present current developments and trends concerning the usage of wastewater resources as a way of growing microalgae for biofuel production purposes.

2. Microalgae

2.1 Background

Microalgae are photosynthetic microorganisms (unicellular or multicellular structure) that can be found in several environments, including marine and freshwater ecosystems. Autotrophic species are able to perform photosynthesis (process similar to that of terrestrial plants), and convert solar energy into chemical energy in the form of biomass (Gonçalves et al. 2014, Kumar et al. 2010). Compared to terrestrial plants, microalgae are more efficient in the use of solar energy to produce biomass (photosynthetic efficiency is about ten times higher), due to the following characteristics (Benemann 1997, Demirbas and Demirbas 2011): simple cell structure; and growth in aqueous environment, which enhances nutrients' mass transfer. These microorganisms were responsible for the current chemical composition of the atmosphere, reducing the CO_2 levels of the primitive atmosphere (Riding 2009, Zavarzin 2005). Even now, they represent one of the most important natural sinks of CO_2. In addition to autotrophic species, some microalgae are heterotrophic or mixotrophic, the last ones being able to intake CO_2 from the atmosphere, or absorb organic compounds from abroad, depending on the conditions in which they are (Brennan and Owende 2010). The biochemical composition of microalgae is essentially a combination of four key elements (Williams and Laurens 2010): carbohydrates, with structural and metabolic function; proteins, which also present structural and metabolic functions, as well as function as biocatalysts (enzymes); nucleic acids, which support the cell division and growth processes; and lipids, which, in addition to the structural function, represent a form of energy storage. Taking into account that the oil productivity of microalgae (oil content multiplied by growth rate) is much higher than other biofuel raw materials, the use of microalgae for biofuel production is theoretically attractive (Pienkos and Darzins 2009, Slade and Bauen 2013, Wijffels and Barbosa 2010).

The classification of microalgal strains depends on the morphological characteristics, the photosynthetic pigments, and the chemical composition of the products of their metabolism. Microalgae can be grouped in several phyla and, in terms of abundance and practical application, the phyla that assume greater prominence are (Hu et al. 2008, Mutanda et al. 2011, Guiry and Guiry 2016): green algae (phylum Chlorophyta), golden algae (phylum Ochrophyta, class Chrysophyceae), and diatoms (phylum Bacillariophyta). Another important group is composed of the prokaryotic cyanobacteria (phylum Cyanobacteria, also known as blue-green algae). These microorganisms also perform photosynthesis, being usually studied in parallel with eukaryotic algae, and generally denominated as microalgae.

2.2 Microalgal Culture Parameters

Microalgal growth is influenced by several parameters (Carvalho et al. 2006, Pires et al. 2012): cultivation pathway; nutrient availability; light; and culture pH and temperature. Cultivation pathway is associated with metabolic needs of microalgae, mainly in terms of supplied energy and carbon source.

2.2.1 Cultivation pathways

In autotrophic culture, microalgae use water and inorganic forms of carbon (CO_2 and dissolved carbonates) to perform photosynthesis. In this type of culture, the presence of contaminations is almost negligible. Thus, it can be performed in either open or closed bioreactors. Light is one of the most important variables in autotrophic cultures. Its intensity should be maintained at saturation level (higher level induces a negative effect on microalgal growth—photoinhibition) during the culturing period. The increase in cell concentration can lead to self-shading—reducing the light supply to the cells in the interior of the bioreactor. The main consequence of light limitation in autotrophic cultures is the low cell densities obtained, when compared to those of heterotrophic cultures. Thus, it requires a high culture volume (and area) to achieve the same biomass productivity. Moreover, the low culture density results in the increase of the downstream process costs, as a large volume of culture should be processed in the harvesting step.

In heterotrophic culture, microalgae use organic forms of carbon (mainly glucose and acetate) as an energy source and as the single carbon source. Thus, light is not required in this type of cultures, simplifying the design and scale-up of the bioreactors (fermenters). The main advantages of this culturing pathway are (Adarme-Vega et al. 2012, Perez-Garcia et al. 2011): the use of the currently available and developed fermentation technology; the high growth kinetics (energy required for metabolic activities is easily processed and not dependent on light); the high cell densities achieved (associated with low harvesting costs); and the high cell lipid content. Consequently, heterotrophic cultivation is considered more economically attractive than autotrophic cultures. However, the composition of the medium increases the probability of culture contamination by bacteria and fungi. Moreover, only few species of microalgae are able to grow heterotrophically.

In mixotrophic culture, microalgae can grow combining the above described nutrition modes (Cheirsilp and Torpee 2012, Heredia-Arroyo et al. 2011, Wan et al. 2011). Some species can also switch between autotrophic and heterotrophic metabolisms, depending on the light and nutrient availability. This flexibility can bring some advantages to microalgal growth. For instance, in autotrophic cultures, the oxygen produced by photosynthesis (if not efficiently removed) can achieve levels that can be toxic to microalgae, causing photooxidative damage (Chisti 2007, Foyer and Shigeoka 2011, Pires et al. 2014). Oxygen utilization through heterotrophy can prevent this phenomenon. In addition, the light availability is not a limitation for biomass production. Consequently, mixotrophic cultures can achieve higher biomass productivities than autotrophic and heterotrophic cultures (Li et al. 2014). Table 2.1 summarizes the main advantages and disadvantages of the three cultivation pathways.

Table 2.1. Comparative features of microalgal cultivation pathways.

Cultivation Pathway	Bioreactor	Advantages	Disadvantages
Autotrophic	Open Ponds	- Temperature rise "control" by evaporation - Low capital costs	- Strong influence of local weather conditions - High probability of contaminations - High requirement of light exposure
	Photobioreactors	- Less loss of water and CO_2 - High photosynthetic efficiency - Higher surface to volume ratio	- Difficulties in scaling-up. - Requirements of temperature control - Requirements of periodic cleaning due to biofilm formation
Heterotrophic	Fermenters	- Control of the optimal conditions for microalgal growth - High biomass concentrations	- High production costs (nutrients) - Competition for feedstocks with other biofuel technologies
Mixotrophic	Photobioreactors	- Low light exposure requirements - High biomass concentration - High lipids productivity	- High maintenance costs - Complex operation - Application to a few microalgal species

2.2.2 *Main required nutrients*

Carbon, nitrogen, and phosphorus are the main nutrients required for microalgal growth. Carbon is essential in both respiratory and photosynthetic metabolism (Sayre 2010, Sydney et al. 2014). In autotrophic cultures, carbon is provided as CO_2 and dissolved carbonates. To enhance biomass concentrations, CO_2-enriched streams should be fed to the culture. Optimal CO_2 concentration depends on microalgal species. There are some species tolerant to very high concentrations; however, for the majority of species, CO_2 becomes toxic above a certain level, mainly due to the decrease of the culture pH (Chinnasamy et al. 2009, Ge et al. 2011, Gonçalves et al. 2016b, Yang and Gao 2003). Besides CO_2 transfer to the cultures, bubbling a gaseous stream in the culture also enhances the mixing (promoting mass and heat transfer), and removes the O_2 produced during photosynthesis.

Nitrogen availability has a strong impact on lipids and fatty acids profiles (Cai et al. 2013). A low concentration of this nutrient in the culture medium (stress conditions) enhances the accumulation of intracellular lipids, which is interesting for biofuel production. However, the lipid productivity decreases due to the low growth rate. Microalgae assimilate inorganic forms of nitrogen from the medium, such as nitrate (NO_3^-), nitrite (NO_2^-) and ammonium (NH_4^+), and convert them into organic nitrogen. In addition, cyanobacteria can also convert atmospheric nitrogen into ammonia by means of fixation (Cai et al. 2013). Microalgae prefer the assimilation of ammonia rather than nitrate, when both inorganic forms are present in the medium. While ammonia is directly assimilated by microalgae, nitrate requires a previous reduction into ammonia (requiring more energy). Consequently, microalgae only consume nitrate when ammonia is almost completely consumed (Cai et al. 2013, Maestrini et al. 1986, Silva et al. 2015).

Phosphorus has an important role in photosynthesis, cell structure and other metabolic activities. This nutrient strongly influences the carbon metabolism through: metabolic energy transfer molecules (adenosine triphosphate, ATP, and nicotinamide adenine nucleotide phosphate, NADPH); intermediates of metabolic pathways (starch and sucrose synthesis/degradation intermediates); and enzyme activity regulation by protein phosphorylation/dephosphorylation. Microalgae prefer the assimilation of the inorganic forms $H_2PO_4^-$ and HPO_4^{2-}. They also have a second mechanism of phosphorus removal from the medium—the luxury uptake. In this mechanism, phosphorus is stored in biomass in the form of polyphosphates (Powell et al. 2009, Powell et al. 2008).

2.2.3 Light

As referred above, light is one of the most important parameters for autotrophic cultures. Light supply influences the synthesis of energetic molecules, carbon flux inside the cells, and the consumption rate of nutrients. Its intensity should be provided in a level that ensures a homogeneous dispersion into the culture, in order to be distributed to all cells. However, an excessive light intensity can cause photoinhibition (Garcia-Camacho et al. 2012, Rubio et al. 2003). Another important parameter that should be taken into account is the photoperiod. If the light intensity is higher than the optimal value, the dark periods in light/dark cycles help microalgae repair any photo-induced damage. The combined effect of light intensity and photoperiod has already been studied for several microalgal species (Gonçalves et al. 2014, Sforza et al. 2012, Wahidin et al. 2013). If light is a controlled variable (artificial light), its intensity should be lower at the beginning of the culture, and should increase along the culturing period, trying to access the algae located in lower layers. On the other hand, the impact of this phenomenon can be reduced by promoting the turbulence of the culture fluid. In this case, the cells will experience short light/dark cycles, which has already been studied (Brindley et al. 2011, Jacob-Lopes et al. 2009, Janssen et al. 2001). Janssen et al. (2001) studied the growth of the marine green alga *Dunaliella tertiolecta* under short light/dark cycles of 3/3 s, 94/94 ms and 31/156 ms (light irradiance was between 440 and 455 $\mu mol\,m^{-2}\,s^{-1}$). The highest biomass productivities were achieved with 94/94 ms cycle, even comparing with the one obtained under continuous light supply. However, turbulence increases the shear stress on cell membrane, whose tolerance depends on microalgal species (cellular death can occur) (Barbosa et al. 2003).

2.2.4 Culture temperature

Temperature is one of the most important culture parameters, influencing microalgal growth rate, cell size, biochemical composition, and consequently nutrient requirements. Optimal temperature for microalgal culture is within the range of 15–26°C. Higher values can inhibit metabolic activity of microalgae, and reduce the CO_2 solubility in the medium. However, some strains (thermo-tolerant) have the ability to grow at higher temperature values, presenting acceptable biomass productivities at 50°C (Bleeke et al. 2014, Ho et al. 2013, Hu et al. 2013). Low temperature values have a negative impact on enzymatic activities associated with photosynthesis (Vonshak and Torzillo 2004).

2.2.5 Culture pH

Regarding pH of the culture, this variable can affect the biochemical processes associated with bio-availability of CO_2 for photosynthesis and the use of nutrient media (Azov 1982, Goldman 1973, Hansen 2002). Due to microalgal CO_2 uptake (in autotrophic cultures) and the low solubility of CO_2 in the culture medium, a significant increase in the pH of the culture can be observed. Thus, this variable should be strictly controlled to ensure that CO_2 is provided in sufficient concentrations for autotrophic growth of microalgae. The pH of the culture should be maintained between 7 and 9 (Pruder and Bolton 1979). For low pH values, the medium can be toxic for microalgae, while an alkaline medium may lead to the precipitation of biomass along with inorganic salts.

3. Nutrient Assimilation Processes

Photosynthesis is one of the most studied natural processes. Basically, the chemical process of extreme importance in photosynthesis is the conversion of CO_2 and water into carbohydrates and O_2. The photosynthetic process is divided in two sequential stages (Keren et al. 2002): the light-dependent reactions; and the light independent reactions. In the first stage, the light is absorbed by chlorophyll, and converted into chemical energy, in the form of the electron carrier molecule NADPH and the energetic storage molecule ATP. In the second stage (also called Calvin cycle), the molecules produced in the previous stage are used in the reduction of CO_2, and in the synthesis of carbohydrates. These have more energy than the sum of the energy present in the used reagents (water and CO_2), which is ultimately explained by transformation and accumulation of light energy (Forti et al. 2003).

Besides CO_2, microalgae may use dissolved inorganic carbon (mainly bicarbonate—HCO_3^-), assimilating it through active transmembrane transport (involving energy expenditure) or by converting it into CO_2 with extracellular enzymatic activity (Sayre 2010, Sydney et al. 2014). Since the conversion is slow, inorganic carbon availability may be a limiting factor for the growth in aqueous media with low CO_2 concentrations (Riebesell and Wolf-Gladrow 2002). To avoid this limitation, many studies have evaluated the possibility of using waste gases of anthropogenic processes (rich in CO_2) as a source of inorganic carbon (Benemann 1997, Pires et al. 2012, Wang et al. 2008).

As the microalgae are not only composed by carbohydrates, they require the assimilation of other nutrients for the production of essential molecules (such as amino acids, nucleic acids and lipids): nitrogen, phosphorus, sulphur, potassium, calcium, magnesium, and chlorine. For normal metabolic functioning, the presence of trace elements (iron, manganese, copper, zinc, cobalt, or molybdenum) and specific vitamins (in the case of microalgal species that do not have the ability to synthesize autonomously) may also be required.

The study of assimilation mechanisms is imperative for the development and optimization of systems that are based on the growth of microalgae. The ratio of nutrient uptake rates can follow the Redfield ratio. This empirical ratio establishes the molecular proportion of carbon, nitrogen and phosphorus that favours photosynthetic aquatic organisms and is defined by C:N:P = 106:16:1 (Redfield 1958). The comparison between this ratio and the one present in the medium enables the identification of limiting nutrients for microalgal growth.

The assimilation of nutrients occurs mainly through molecular diffusion. This phenomenon is dependent on three factors: the concentration gradient between the medium and the interior of the cells; the nutrient diffusion coefficient; and the diffusive boundary layer thickness (Riebesell and Wolf-Gladrow 2002). Nitrogen can be absorbed in the forms of ammonium-nitrogen (NH_4-N) and nitrate-nitrogen (NO_3-N), whereas phosphorus is incorporated in the form of phosphate-phosphorus (PO_4-P). Similarly, sulphur can be assimilated as sulphate-sulphur (SO_4-S), being important in the formation of essential amino acids (Giordano and Prioretti 2016).

Potassium is a crucial element in osmotic regulation, protein synthesis and as a co-factor in various enzymes (Iyer et al. 2015). It is assimilated under the K^+ ion form.

3.1 Microalgal Cultivation Bioreactors

Microalgae can grow in a wide range of habitats. Thus, their cultivation can be performed in both open and closed systems. The selection of the cultivation system is dependent on several factors (Borowitzka 1999, Eriksen 2008, Janssen et al. 2003, Posten 2009, Richardson et al. 2014): capital and operational costs; value of the target products; cultivation pathway; CO_2 source (in autotrophic cultures); light supply (from natural or artificial sources); and available nutrient sources.

3.1.1 Open ponds

Open ponds are the most commonly used cultivation systems at a commercial scale, due to the low capital and operational costs (Borowitzka 1999, Eriksen 2008). An important drawback is the high probability of contamination with: local algae or other organisms (bacteria and fungi); and dust particles, leaves, and other airborne materials. This phenomenon causes variability in the achieved biomass quality and productivity. To reduce the contamination effect, severe environments are selected to favour the growth of the selected microalgal species, such as *Dunaliella, Spirulina/Arthrospira*, and *Chlorella*, that are cultivated in media with high salinity, alkalinity, and nutrition, respectively (Juneja et al. 2013). Moreover, the culture in this type of bioreactor is strongly affected by the local weather conditions. Water loss by evaporation and low efficiency in CO_2 mass transfer from gaseous to liquid phases are other drawbacks. Regarding the use of light, photosynthetic efficiency is usually below 3%, which is lower than the value achieved in closed photobioreactors (6.5%) (Norsker et al. 2011).

3.1.2 Closed photobioreactors

Closed systems have attracted the interest of several researchers, because they allow the control of nearly all biotechnological parameters, while reducing contamination risks and CO_2 and water losses (Borowitzka 1999, Eriksen 2008). Additionally, it is also easier to obtain reproducible cultivation conditions. High photosynthetic efficiency and, consequently, high biomass productivity are two important advantages, especially when biomass is intended to be used on pharmaceutical or food industries (Cheah et al. 2015). In the last years, the design of closed bioreactors has been improved to make better use of the control parameters that may beneficially affect microalgal growth. There are several photobioreactor designs (Janssen et al. 2003, Posten 2009): air-lift; bubble column; and flat plate.

Airlift and bubble column photobioreactors are the most well-known closed systems. These bioreactors have wide acceptance for algal cultivation due to their simplicity in gas-liquid contacting application. Airlift photobioreactor is an interesting bioreactor because it promotes high mass transfer rates, regular light/dark cycles, and low and homogeneous shear stress to the cells. This bioreactor is a vessel with two interconnecting zones: one of the tubes is the riser where gas mixture is sparged, whereas the other region is called downcomer. Usually, it presents two main designs (Janssen et al. 2003, Pires et al. 2012, Posten 2009): in the internal loop, the riser and downcomer are separated either by a draft tube or a split-cylinder; and in the external loop, these regions are separated physically into different splits. Mixing is done by bubbling the gas through the sparger in the riser tube without any physical agitation. On the other hand, bubble column photobioreactor permits to obtain high process efficiencies with lower energy consumption, higher mass transfer rates, well known flow patterns, and shorter circulation times. Mixing and CO_2 mass transfer are also ensured through bubbling the gas mixture from sparger. Both photobioreactor designs are widely used in industrial processes, as they present several advantages: low power consumption; high mass transfer capacity; and high photosynthetic efficiency.

Flat plate (or panel) photobioreactor is a cuboidal shape characterized by high surface area to volume ratio, and open gas disengagement systems (Janssen et al. 2003, Posten 2009). Mixing is performed with a mechanical pump, either by bubbling air from its one side through a perforated tube, or by rotating it mechanically through a motor. Although this type of bioreactor is cost effective and has excellent biomass yields, it presents some limitations, such as difficulties in temperature control, limited degree of growth in the region close to the wall, and hydrodynamic stress.

4. Biomass Applications

The use of microalgae as a raw material for various processes has recently increased, given the need to find replacements for the most common biomass sources. The algae growth presents several advantages over growing plants commonly used in the area of biorefinery. Biomass and lipids productivities per unit area can be between 7 to 31 times higher than those achieved by the terrestrial plant presenting the highest growth rate: the palm oil tree (Chisti 2007). Consequently, microalgae are the most promising and sustainable source for biofuel production.

Biofuel production follows one of the following pathways: thermochemistry or biochemistry. Regarding thermochemistry, the biomass is subjected to processes with different temperature ranges, in order to obtain products with distinct characteristics. It is possible to obtain: gas by gasification; bio-oil by liquefaction; a mixture of products by pyrolysis; or electricity by direct combustion. The alternative to these processes uses the biological activity of certain bodies to produce more specific fuels. For instance, through anaerobic digestion, a mixture of methane (CH_4) and hydrogen (H_2) can be obtained (usually known as biogas). This mixture typically presents a high calorific value, making it suitable for power generation. Another potential process is the production of ethanol through alcoholic fermentation that has been applied for other raw materials. The produced ethanol can be used as fuel additive for application in the propulsion of motor vehicles (Brennan and Owende 2010).

The biorefinery concept is defined as the sustainable biomass conversion into a spectrum of bio-products (food, feed, and chemicals) and bioenergy (biofuels and heat). The biorefinery is an approach that allows the obtainment of the maximum return from the biomass components, converting them into marketable products, and thus contributing to the development of more economically and environmentally sustainable solutions (Clark et al. 2012). Regarding microalgae, the interest in biofuel production is associated with its high biomass and lipid productivities, but several other constituents have high commercial interest. The most studied microalgal biofuels are biodiesel, bioethanol, and biomethane (Singh and Cu 2010). Producing biodiesel from microalgae presents a huge potential, since it can be used in common propulsion engines without additional technological modifications.

The production of microalgal biomass is not limited to the generation of biofuels; other chemical compounds of commercial interest can be generated from this. The microalgal biomass can be used for the production of nutritional supplements, animal feed, and aquaculture, as well as other chemical compounds with commercial value, such as cosmetics and pharmaceuticals, fertilizer for soils, and colouring agents (Pulz and Gross 2004, Raja et al. 2008). All applications are based on products of the metabolism of microalgae, i.e., biomass, carotenoids and antioxidants, fatty acids, enzymes, polymers, and other specific biochemical compounds used in scientific research.

Taking into account all the potential provided by microalgae, its commercial exploitation can be economically viable. Therefore, the integration of processes and optimization of the revenues achieved with microalgal biomass are the key issues to achieve an economically viable and sustainable biofuel production using microalgae.

5. Wastewater as Microalgal Culture Medium

5.1 Wastewater Treatment

The increase in urbanization and industrialization and the improvement of the quality of life have raised new environmental challenges, mainly in the management of generated wastes (solids, liquids, or gases). The socio-economic development is also associated with increasing efforts to improve the health conditions of populations, and to minimize the environmental impact of industries and anthropogenic activities. Consequently, there is a need to improve and renew the waste treatment technologies. Wastewater, as any other waste, is an unwanted by-product that requires the implementation of purification systems capable of changing its characteristics, so that it can be safely discharged (with reduced environmental impacts). The development of these treatment systems must be done taking into account the origin and composition of the effluent to be treated, and the sensitivity of aquatic ecosystem that will receive it.

Wastewater treatment technologies have been developed to respond to the problems associated with effluents' discharge in water resources, naturally sensitive to any variations in their compositions (Tchobanoglous et al. 2003). Based on physical forces, chemical reactions, and biological mechanisms, they are intended to give a safe character to water, through the modification and adaptation of its physicochemical and biological characteristics. Traditionally, a wastewater treatment plant (WWTP) is composed of various processes that can be organized in different treatment stages: preliminary, primary, secondary, and tertiary (or advanced).

5.1.1 Preliminary treatment

The effluent to be treated may contain several kinds of materials. Preliminary processes aim the removal of solids with large dimensions, which can cause problems once they reach some of the downstream treatment steps, including the wear of pumps and pipes (Davis 2011). Typically, the first treatment step is the railing; this is the obstruction of floating solids pass through a grid system with a given spacing that sets the maximum value for the minimum size of objects that can pass. Then, it is common to use tanks to promote the gravitational sedimentation of fine sands.

5.1.2 Primary treatment

The primary treatment has as its main goal the removal of organic and inorganic solids liable to sedimentation (oils and fats and colloidal matter) using physicochemical phenomena (Davis 2011). This step aims to remove settleable organic and floatable solids (removal efficiencies of 90 to 95% settleable solids, 40 to 60% of total suspended solids—TSS, and 25 to 35% of biochemical oxygen demand—BOD—are expected) (Spellman 2013). The removal of some metallic species, as well as nitrogen and phosphorus (present in suspended particles) can also occur.

The main process of primary treatment is sedimentation, often being referred to as primary sedimentation. This step can be preceded by a neutralization step, through mechanical mixing and/or addition of chemicals. The sedimentation of suspended matter can be made through primary decanters with retention times of two to three hours and with several possible configurations. The removal of oils and fats can be made superficially using scrapers, and can be enhanced and induced through water pressurization or decompression mechanisms. This step can be adapted and incorporated into the preliminary treatment. A process of coagulation and flocculation is also possible at this stage. By the addition of certain substances, it is possible to form nuclei of colloidal matter and neutralize their loads (coagulation) and particles large enough to be separable by the action of gravity forces (flocculation). Finally, phosphorus removal can also be promoted in this treatment phase, by adding metallic salts and inducing the precipitation of phosphates.

5.1.3 Secondary treatment

Secondary treatment processes have emerged as a simulation of processes observed in natural systems and carried out by various microorganisms (Davis 2011). For this reason, they may also be known as biological treatment processes. Metabolic pathways of one or more microbial species are used for the transfer of soluble-phase organic pollutants to suspended matter (in the form of biomass). Usually aerobic processes strongly depend on external supply of oxygen. However, they can also use anaerobic or anoxic processes to promote the removal of some nutrients or convert them into species that can be used by aerobic microorganisms. The most commonly used aerobic process is the activated sludge, which can be implemented in several variants and with different results (Spellman 2013). Secondary treatment step is defined by Clean Water Act as the process that enables the achievement of an effluent with no more than 30 mg L^{-1} BOD and 30 mg L^{-1} TSS. Secondary treatment processes can be divided in two groups: fixed films systems and suspended growth systems. In the first group, microorganisms are attached to a surface. On the other hand, in suspended growth systems, they grow within the wastewater environment.

5.1.4 Tertiary/advanced treatment

Tertiary or advanced treatment of wastewater is typically used when the preceding steps are not capable of producing an effluent with features that allow their discharge in water resources. Advanced treatment options are diverse and are intended for specific cases. Among all, the most used are the chlorination, ozonation and exposure to ultraviolet radiation (with the goal of chemical disinfection), and filtering (for removal of suspended solids in excess). Depending on the pollutants to remove or, if so apply, depending on the purpose of the treated water, more specific processes, such as adsorption by activated carbon, reverse osmosis, ion exchange, and electro-dialysis can be adopted.

5.2 Wastewater Issues and Current Solutions

Although common treatment systems are efficient in removing most of the pollution load of the effluent, some phenomena can occur, significantly affecting its composition and putting into question the treatment efficiency. To solve this type of problem, it is necessary to develop specific solutions that complement the treatment steps described in the previous section.

5.2.1 High nutrient concentrations

One of the problems associated with the inefficiency of currently applied treatment steps is the presence of excessive nutrients concentrations in the wastewater to be treated. In these cases, if the treatment is not efficient, the resulting effluent will significantly affect the balance of aquatic ecosystems where it is discharged. This consequence is traditionally known as eutrophication, and has proven to be a very adverse and hard-to-control phenomenon that negatively impacts media quality (Khan and Mohammad 2014). Eutrophication caused by discharge of treated wastewater may be preventable by changing the nutritional characteristics of the tributary system, or by changing the system itself (by the addition or adaptation of a specific treatment). The early eutrophication, properly controlled with specific upstream biological processes, can significantly reduce the levels of nutrient concentrations. This process will be discussed further on. It should be noted that the operation and control of this type of biological processes are specialized tasks and highly sensitive to variations in the quality of the effluent, which logically leads to operational and financial efforts.

5.2.2 Heavy metals

Another emerging problem with high importance to the scientific community is the appearance of heavy metals, such as arsenic, lead, and mercury, in wastewaters (including industrial) and in landfill leachates. Although some species are essential micronutrients (in very small concentrations), they can have acute or chronic toxic effects (by metabolic interference or mutagenesis) to organisms, when present in high concentrations (Buruiana et al. 2015). This toxicity is a problem for both the media receivers and to biological-based treatment systems, since these can be severely affected. Currently, the main technologies that allow the removal of these species from the waters present high costs, due to the complex physical or chemical processes involved, and the use of expensive reagents. The most common options include chemical precipitation, ion exchange, membrane filtration, adsorption mechanisms, coagulation-flocculation, or floating and electrochemical methods (Fu and Wang 2011).

5.2.3 Micropollutants

Micropollutants are individual compounds or complex mixtures resulting from anthropogenic activities that appear in very small concentrations (in the order of micrograms per litre) in natural waters (Eggen et al. 2014). Some examples of such substances include pharmaceuticals, pesticides and solvents. Their exact effects on ecosystems are still unknown within the scientific community, since they belong to a class of relatively new pollutants. However, the fact that they are not completely biodegradable, and that the conventional treatment technologies are not sufficient for their removal, makes them emerging pollutants that need to be taken into account. Currently, the solutions which are more efficient in their treatment are the oxidative treatment with ozone, and the activated charcoal powder, both presenting high costs (Eggen et al. 2014).

5.3 Microalgal Application to Wastewater Treatment

Due to their characteristics, microalgae can be applied in different steps of wastewater treatment. In the secondary treatment, autotrophic microalgae can be applied to absorb minerals oxidized by native bacteria and, at the same time, enrich the water with oxygen to promote an aerobic environment (Covarrubias et al. 2012, Gonçalves et al. 2016a, Pires et al. 2013). This consortium between heterotrophic bacteria and autotrophic microalgae substantially reduces the requirements for wastewater aeration, one of the greatest costs in the treatment process. Consortia between bacteria and microalgae, and between different microalgal species have already been studied for wastewater treatment (De-Bashan et al. 2008, Hernandez et al. 2009, Pires et al. 2013).

Within the tertiary treatment step, microalgae are mainly studied to remove nutrients (nitrogen and phosphorus) from wastewater (Orpez et al. 2009, Silva et al. 2015). They present high efficiency in the removal of nutrients, because these nutrients are essential for both autotrophic and heterotrophic metabolic processes. As previously referred, carbon is also important to achieve high growth rates and consequently high nutrients removal rates: the presence of organic carbon can have great influence particularly in heterotrophic cultures, while CO_2 and dissolved carbonates are important for autotrophic cultures. Nitrogen and phosphorus are two elements that are part of the composition of wastewaters from different sources. Table 2.2 presents the generic composition of wastewaters from different sectors. Pittman et al. (2011) analysed in greater detail the characteristics and the suitability of each category of wastewater, with regard to their potential for biofuels production. The current challenge goes through optimizing the process, identifying the ideal concentrations of nitrogen and phosphorus for different culture conditions. The N/P ratio seems to be essential, since studies with high concentrations of these elements and low amounts of others, demonstrated a reduction in the growth rate (Li et al. 2010). The so-called Redfield ratio (106C:16N:1P),

Table 2.2. Composition of wastewaters in terms of total nitrogen (TN) and total phosphorus (TP) (adapted from Cai et al. (2013) and Henze and Comeau (2008)).

Wastewater	Source	TN (mg L⁻¹)	TP (mg L⁻¹)
Municipal	Sewage	15–90	5–20
Industrial and agro-industrial	Dairy	185–2,636	30–727
	Poultry	802–1,825	50–446
	Swine	1,110–3,213	310–987
	Beef feedlot	63–4,165	14–1,195
	Textile	21–57	1.0–9.7
	Winery	110	52
	Tannery	273	21
	Paper mill	1.1–10.9	0.6–5.8
	Olive mill	532	182
Anaerobic Digestion	Dairy manure	125–3,456	18–250
	Poultry manure	1,380–1,580	370–382
	Sewage sludge	427–467	134–321
	Food waste	1,640–1,885	296–302
Leachate	Landfill	100–500	1–10

which estimates the elemental composition of microalgal cells, is used as a starting point for estimating the N/P ratio in the medium. It is assumed that the critical ratio will be 16:1. However, several studies have tested different ratios. Recent studies have estimated that the ideal ratio for the growth of microalgae is expected to be around 7–8 (Cai et al. 2013, Li et al. 2010). Therefore, it appears that the combination of microalgal production with wastewater treatment may be a viable alternative to reduce process costs.

Several authors have already tested different species of microalgae for the treatment of effluents with different concentrations of nitrogen and phosphorus, in order to verify their efficiency in the removal process. The optimization of the N/P ratio is a key to the profitability of the process, since the limitation of nitrogen can lead to a lower growth rate. Therefore, more studies on this parameter are needed, to maximize biomass production from wastewaters. The study of the kinetics of the removal of nutrients is important to better understand how they are assimilated by microalgae. This information is essential for a correct dimensioning of processes on a larger scale, since it is necessary to know how fast the assimilation of different nutrients is to calculate the retention time in ponds.

In addition to the assimilation of nutrients, another feature of microalgae is their ability to perform biosorption. This can be used in the remediation of contaminated media, and in the accumulation and potential reuse of some substances, such as heavy metals. This process allows the concentration of certain substances in cell structure, without power consumption (Volesky and Holan 1995). This ability of microalgae is comparable to chemical adsorbents, meaning that microalgae can be a viable alternative method of metals' removal and recovery (Mehta and Gaur 2005). The microalgal removal capacity of micropollutants in wastewaters has also been reported in the literature (de Wilt et al. 2016, Hirooka et al. 2003, Ji et al. 2014). These studies have demonstrated that microalgae constitute good remediation agents of water contaminated with pharmaceutical substances and various organic pollutants, which could prove another advantage of their use in the treatment of effluents.

5.3.1 Case studies

Microalgal cultures have already been performed in several real wastewater effluents. The most commonly studied effluent is domestic/municipal wastewater, because it does not present toxic compounds that inhibit microalgal growth, and contains a N/P ratio close to the theoretically ideal (Rawat et al. 2011). On the other hand, the wastewater from agriculture shows nitrogen and phosphorus levels higher than those found in municipal effluents. Some studies have proven that growth and consequent production of biomass in these media can even be superior, given the surplus of nutrients. However, this wastewater can also contain pesticides and herbicides, inhibiting microalgal growth (Cai et al. 2013). Wastewater from industries presents quite different compositions, depending on the commercial sector. The main interest in the application of microalgae in the treatment of this effluent type is the removal of heavy metals (cadmium, chromium, zinc, etc.) and other toxic organic compounds. However, the low concentration of nitrogen and phosphorus results in relatively lower growth rates and biomass productivities, as compared with those obtained using other types of wastewater (Pittman et al. 2011). Anaerobically-digested effluents, typically present large amounts of nitrogen and phosphorus, being potential candidates to support microalgal growth. Leachate is a big problem for municipal solid waste landfills, and constitutes a significant threat to surface and ground waters. It results from the precipitation, in which the water passes through landfills, dissolving suspended matter from it. Landfill leachate is a difficult treatment effluent, since its composition floats a lot, depending on the stage of decomposition of the waste and on the origin of the components of the mixture. In any case, nutrient concentration profile may be not favourable for microalgal growth: the concentration of NH_4-N varies between 50 and 2,200 mg N L^{-1} and the range of TP concentrations goes from 0.1 to 23 mg L^{-1} (Kjeldsen et al. 2002).

5.3.2 *Municipal wastewaters*

Municipal wastewaters have been widely used in microalgal cultivation for nutrients removal and for other purposes. Table 2.3 presents some examples where microalgae have been effectively applied in nutrients removal from municipal wastewaters.

   Although the majority of the studies refer to the use of suspended-growth systems, Shi et al. (2007) have assessed nitrogen and phosphorus removal from a municipal wastewater collected in Cologne (Germany) using an immobilization method—the twin-layer system. In this method, the microalgae *Chlorella vulgaris* and *Halochlorella rubescens* (formerly *Scenedesmus rubescens*) were immobilized through self-adhesion on a substrate layer, and another layer provided the growth medium required for microalgal growth. Using this system, microalgae remained 100% immobilized, being able to completely remove NO_3-N (initial concentration between 3.7–6.2 mg N L^{-1}) after an exposure period of four days. Similarly, Whitton et al. (2016) used an immobilization technique to evaluate the influence of microalgal biochemical composition on the removal of nutrients from a municipal wastewater collected in a WWTP located in the Midlands (UK). In this study, the authors immobilized the microalgae *Tetradesmus obliquus* (formerly *Scenedesmus obliquus*) and *C. vulgaris* in calcium-alginate beads, obtaining NH_4-N removal rates ranging between 0.19 and 0.50 mg N

Table 2.3. Application of microalgae in nitrogen and phosphorus removal from municipal wastewaters and respective removal rates.

Microorganisms	Nitrogen		Phosphorus		Reference
	C_i (mg N L^{-1})	RR (mg N L^{-1} d^{-1})	C_i (mg P L^{-1})	RR (mg P L^{-1} d^{-1})	
Chlorella vulgaris/ Tetradesmus obliquus (formerly *Scenedesmus rubescens*)	3.7–6.2 NO_3-N	0.90–1.52	0.4–0.7 TP	n.a.	(Shi et al. 2007)
Chlamydomonas reinhardtii	128.6 TN	3.44	120.6 TP	0.56	(Kong et al. 2009)
Chlorella sp.	16.95 NO_3-N	1.18	0.32 TP	≈ 0	(Wang et al. 2010)
Chlorella sp./ *Chlorella vulgaris/ Scenedesmus communis* (formerly *Scenedesmus quadricauda*)/ *Tetradesmus dimorphus* (formerly *Scenedesmus dimorphus*)	70 NO_3-N	10–12.7	16 PO_4-P	3.83–5	(Singh and Thomas 2012)
Native microalgae	50.1 TN	3.18	8.8 PO_4-P	0.34	(Su et al. 2012)
Chlorella vulgaris	18.3 TN	3.05	0.5 PO_4-P	0.07	(Ruiz et al. 2013)
Parachlorella kessleri (formerly *Chlorella kessleri*)/ *Chlorella vulgaris/ Nannochloropsis oculata*	130 TN	11.36	5.76 TP	0.52	(Caporgno et al. 2015)
Chlorella vulgaris	7.1–10.4 TN	6.75–9.88	0.23–1.84 TP	0.18–1.47	(Filippino et al. 2015)
Chlorella vulgaris/ Chlorella sp./ *Scenedesmus fuscus* (formerly *Chlorella fusca*)/ *Desmodesmus subspicatus* (formerly *Scenedesmus subspicatus*)/ *Raphidocelis subcapitata* (formerly *Pseudokirchneriella subcapitata*)	17.2 TN	0.57	10 TP	0.33	(Gómez-Serrano et al. 2015)

Table 2.3 cont. ...

... Table 2.3 cont.

Microorganisms	Nitrogen		Phosphorus		Reference
	C_i (mg N L^{-1})	RR (mg N L^{-1} d^{-1})	C_i (mg P L^{-1})	RR (mg P L^{-1} d^{-1})	
Ankistrodesmus falcatus/Tetradesmus obliquus (formerly *Scenedesmus obliquus*)/ *Parachlorella kessleri* (formerly *Chlorella kessleri*)/*Chlorella vulgaris/Chlorella sorokiniana/ Botryococcus braunii/ Ettlia oleoabundans* (formerly *Neochloris oleoabundans*)/ Native microalgae	54.58 TN	13.49–27.2	12.70 TP	1.2–11.7	(Mennaa et al. 2015)
Chlorella vulgaris / Tetradesmus obliquus (formerly *Scenedesmus obliquus*)/Native microalgae	119.3–346.6 NH$_4$-N	6.28–25.85	4.6–8.3 PO$_4$-P	0.32–0.41	(Gouveia et al. 2016)
Scenedesmus sp.	38.6 NH$_4$-N 17.1 NO$_3$-N	5.16–5.42 1.38–1.71	9.24 PO$_4$-P	0.96–1.08	(Nayak et al. 2016)
Tetradesmus obliquus (formerly *Scenedesmus obliquus*)/*Chlorella vulgaris*	5 NH$_4$-N	0.19–0.50	0.5–10 PO$_4$-P	0.009–0.82	(Whitton et al. 2016)

C_i—initial concentration (mg L^{-1}); RR—removal rate (mg L^{-1} d^{-1}).

L^{-1} d^{-1} and PO$_4$-P removal rates ranging between 0.009 and 0.82 mg P L^{-1} d^{-1}. Additionally, the authors have concluded that higher removal efficiencies can be achieved by species presenting high N/P internal ratios.

Other studies have focused on the use of municipal wastewaters to improve microalgal biomass and lipid productivities. For example, Kong et al. (2009) have grown *Chlamydomonas reinhardtii* in a municipal wastewater obtained from St. Paul Metro plant (St. Paul, Minnesota, USA), achieving nutrients removal rates of 3.44 mg N L^{-1} d^{-1} and 0.56 mg P L^{-1} d^{-1} for TN and PO$_4$-P, respectively. Using a municipal wastewater, average biomass productivities achieved by *C. reinhardtii* ranged between 0.82 and 2.00 g L^{-1} d^{-1}, with 25.25% (w/w) of microalgal biomass corresponding to lipids. Similarly, Caporgno et al. (2015) have grown *Parachlorella kessleri* (formerly *Chlorella kessleri*), *C. vulgaris* and *Ettlia oleoabundans* (formerly *Nannochloropsis oculata*) on a municipal wastewater obtained from the MWTP of Saint

Nazaire (France) to evaluate their potential in nutrients removal, and biomass and bioenergy (biodiesel and methane) production. With this study, high biomass concentrations and removal efficiencies of nutrients were achieved by the studied microalgae: maximum biomass concentrations achieved ranged between 2.70 and 2.91 g L⁻¹ and removal efficiencies of nutrients ranged between 95 and 96% for nitrogen, and between 98 and 99% for phosphorus. Additionally, this study has demonstrated the potential of microalgal cultures grown in wastewaters for biofuels production, since the studied microalgae were able to produce between 346 and 415 mL CH_4 g⁻¹ biomass during anaerobic digestion and between 74 and 113 mg biodiesel g⁻¹ biomass. Gómez-Serrano et al. (2015) have also demonstrated that coupling microalgal biomass production to the tertiary treatment of wastewaters promotes recovery of nutrients, while contributing to the production of valuable biomass. In this study, the authors have grown *Muriellopsis* sp., *C. vulgaris*, *Scenedesmus fuscus* (formerly *Chlorella fusca*), *Chlorella* sp., *Desmodesmus subspicatus* (formerly *scenedesmus subspicatus*) and *Raphidocelis subcapitata* (formerly *Pseudokirchneriella subcapitata*) in a municipal wastewater collected in Almeria (Spain) trying to simulate outdoor conditions. The authors have concluded that all the microorganisms that were studied were able to effectively remove nitrogen and phosphorus present in the wastewater and, at the same time, accumulate large amounts of lipids and carbohydrates (fatty acids content was about 25% w/w, which corresponds to a productivity of 110 mg L⁻¹ d⁻¹). Additionally, the authors have proposed *Muriellopsis* sp. and *Desmodesmus subspicatus* as promising microalgae for outdoor production using municipal wastewaters as culture medium, since these microalgae presented the highest biomass productivities and photosynthetic efficiencies, as well as the highest nutrient coefficient yields. Mennaa et al. (2015) have studied the potential of seven microalgal species [(*Ankistrodesmus falcatus*, *Tetradesmus obliquus* (formerly *S. obliquus*), *Parachlorella kessleri* (formerly *C. kessleri*), *C. vulgaris*, *Chlorella sorokiniana*, *Botryococcus braunii*, and *Neochloris oleoabundans*)], and a natural bloom on biomass production, and removal of nitrogen and phosphorus from a municipal wastewater supplied by a MWTP from the south of Spain. From the microorganisms studied, the authors have concluded that the most effective in both biomass production and nutrients removal were *T. obliquus* and the natural bloom, with biomass productivities ranging between 108 and 118 mg L⁻¹ d⁻¹ and nitrogen and phosphorus removal efficiencies of more than 80 and 87%, respectively. More recently, Gouveia et al. (2016) have cultured *C. vulgaris*, *T. Obliquus*, and a native consortium in a municipal wastewater collected from Figueira da Foz (Portugal), aiming to determine the best candidate in terms of wastewater remediation and biomass productivity and quality for further uses, such as biofuels, biofertilizers and bioplastics production. With this study, biomass productivities achieved were 0.1, 0.4 and 0.9 g L^{-1} d^{-1} for *C. vulgaris, T. obliquus* and the consortium, respectively, with 8.1, 10.0 and 13.1% (w/w) corresponding to lipids. Additionally, the studied cultures have effectively removed nitrogen and phosphorus from the wastewater, reaching nitrogen removal efficiencies of 84–98% and phosphorus removal efficiencies of 95–100%. Taking into account these results, the authors have proposed the native consortium as the best option for the removal of nutrients and biomass production. Nayak et al. (2016) have grown different microalgal species [(*C. vulgaris, Mychonastes homosphaera* (formerly *Chlorella minutissima*), *Scenedesmus* sp., and *Chlorococcum* sp.)] in a municipal wastewater collected in Kharagpur (India) to evaluate their potential in simultaneous wastewater treatment, CO$_2$ uptake, and lipid biosynthesis for biofuels production. This study has demonstrated that *Scenedesmus* sp. has proven to be the most effective in terms of biomass production (biomass productivities determined for this microalga were 61.4 mg L^{-1} d^{-1}) and lipid biosynthesis (lipid content was about 23.1% w/w). Taking into account these results, further studies were carried out using this microalga. Accordingly, when supplemented with 2.5% (v/v) CO$_2$, *Scenedesmus* sp. cultures were able to achieve a biomass productivity of 196 mg L^{-1} d^{-1}, lipid content of 33.3% (w/w), which corresponds to a lipid productivity of 65.17 mg L^{-1} d^{-1} and CO$_2$ uptake rate of 368 mg L^{-1} d^{-1} (Nayak et al. 2016).

5.3.3 Industrial/agro-industrial wastewaters

Wastewaters resulting from industrial and agro-industrial practices have also been used for microalgal growth with different purposes: removal of nutrients; biomass production; and bioenergy production. As it is possible to observe in Table 2.4, industrial and agro-industrial wastewaters from different sources have been used for these purposes.

Regarding the removal of nutrients, Valderrama et al. (2002) have cultured *C. vulgaris* in an industrial effluent resulting from ethanol and citric acid production, achieving NH$_4$-N and PO$_4$-P removal efficiencies of 71.6 and 28%, respectively (initial NH$_4$-N concentration in this effluent ranged between 3 and 8 mg N L^{-1}, whereas initial PO$_4$-P concentration ranged between 0 and 0.36 mg P L^{-1}). Similarly, Lim et al. (2010) have grown *C. vulgaris* in high rate algal ponds fed with a textile industry wastewater to evaluate the potential of this microalga in nitrogen and phosphorus removal. Although *C. vulgaris* was able to grow in the textile wastewater (NH$_4$-N and PO$_4$-P initial concentrations of 6.50 and 7.14 mg L^{-1}, respectively), nitrogen and phosphorus removal efficiencies achieved were not very high: 44.4–45.1% and 33.1–33.3%, respectively. In the study performed by Hernández et al. (2013), *C. sorokiniana* was grown in a potato-processing wastewater presenting initial NH$_4$-N concentration of 12.1 mg

Table 2.4. Application of microalgae in nitrogen and phosphorus removal from industrial and agro-industrial wastewaters and respective removal rates.

Wastewater source	Microorganisms	Nitrogen		Phosphorus		Reference
		C_i (mg N L^{-1})	RR (mg N L^{-1} d^{-1})	C_i (mg P L^{-1})	RR (mg P L^{-1} d^{-1})	
Ethanol and citric acid industry wastewater	*Chlorella vulgaris*	3–8 NH_4-N	0.36– 1.25	1.5– 3.5 PO_4-P	0–0.36	(Valderrama et al. 2002)
Piggery wastewater	*Botryococcus braunii*	788 NO_3-N	58.7	40 PO_4-P		(An et al. 2003)
Rice culture wastewater	*Aphanothece microscopica* Nägeli	48.43 TN	56.4	52.82 PO_4-P	n.a.	(Queiroz et al. 2007)
Carpet mill industry wastewater	Native microalgae	1.4–3.9 NO_3-N	0.47– 1.30	17.6– 22.0 PO_4-P	5.67– 7.27	(Chinnasamy et al. 2010)
Textile industry wastewater	*Chlorella vulgaris*	6.50 NH_4-N	0.24	7.14 PO_4-P	0.20	(Lim et al. 2010)
Wastewater treatment plant from an industrial park in Taiwan	*Chlamydomonas* sp.	38.4 NH_4-N 3.1 NO_3-N	3.84 0.31	44.7 PO_4-P	1.48	(Wu et al. 2012)
Potato-processing wastewater	*Chlorella sorokiniana*	12.1 NH_4-N	1.15	3.4 PO_4-P	0.27	(Hernández et al. 2013)
Piggery wastewater	*Chromochloris zofingiensis* (formerly *Chlorella zofingiensis*)	148.0 TN	11.65	156.0 TP	13.26	(Zhu et al. 2013)
Aquaculture wastewater	Microalgal-bacterial flocs	33.7 TN	0.98	3.52 TP	0.16	(Van Den Hende et al. 2014)
Chemical industry wastewater	Microalgal-bacterial flocs	80.2 TN	1.34	3.31 TP	0.02	(Van Den Hende et al. 2014)
Food-processing industry wastewater	Microalgal-bacterial flocs	128 TN	2.17	12.3 TP	0.20	(Van Den Hende et al. 2014)

Table 2.4 cont. ...

... Table 2.4 cont.

Wastewater source	Microorganisms	Nitrogen		Phosphorus		Reference
		C_i (mg N L^{-1})	RR (mg N $L^{-1} d^{-1}$)	C_i (mg P L^{-1})	RR (mg P $L^{-1} d^{-1}$)	
Manure treatment industry wastewater	Microalgal-bacterial flocs	46.2 TN	0.46	0.41 TP	0.01	(Van Den Hende et al. 2014)
Meat-processing industry wastewater	Native microalgae	13.4 NH_4-N	2.01	3.6 PO_4-P	0.25	(Assemany et al. 2016)
Horticultural wastewater	*Klebsormidium* sp./ *Stigeoclonium* spp.	47.2 NO_3-N	3.5–6.4	11.6 PO_4-P	3.3–3.9	(Liu et al. 2016)

C_i—initial concentration (mg L^{-1}); RR—removal rate (mg $L^{-1} d^{-1}$).

N L^{-1}, and PO_4-P concentration of 3.4 mg P L^{-1}. After a cultivation period of 10 days, nitrogen and phosphorus removal efficiencies achieved were 95 and 80.7%, respectively. Van Den Hende et al. (2014) have applied the microalgal-bacterial flocs technology to promote nutrients removal from four different industrial wastewaters (aquaculture, chemical industry, food-processing industry, and manure treatment industry wastewaters). This technology is based on microalgal-bacterial interactions, where mechanical aeration can be replaced by photosynthetic aeration, and on flocs formation by the microorganisms involved, which significantly improves microalgal separation from the treated wastewater. Using this technology, the authors have reported the following nitrogen and phosphorus removal efficiencies, respectively: 57.9 and 88.6% for aquaculture wastewater; 45.1 and 18.7% for chemical industry wastewater; 45.9 and 44.6% for food-processing industry wastewater; and 19.7 and 53.7% for manure treatment industry wastewater. Liu et al. (2016) have determined the remediation potential of the filamentous microalgae *Klebsormidium* sp. and *Stigeoclonium* spp. grown in an outdoor Algal Turf Scrubber using horticultural wastewater as culture medium (NO_3-N and PO_4-P initial concentrations of 47.2 and 11.6 mg L^{-1}, respectively). With this study, the authors have demonstrated that these microalgae can effectively remove nitrogen and phosphorus from this wastewater, since nitrogen removal efficiencies achieved oscillated between 88–99%, and phosphorus removal efficiencies were higher than 99%.

Studies combining removal of nutrients and bioenergy production have also been reported in the literature for industrial and agro-industrial wastewaters. For example, Zhu et al. (2013) have grown *Chromochloris zofingiensis* (formerly *Chlorella zofingiensis*) in a piggery wastewater to assess its potential on nutrients removal and biodiesel productivity. After 10 days of culturing, TN and TP removal efficiencies achieved in this study were 78.7 and 85%, respectively. Regarding oil content determined in *C. zofingiensis* grown in the piggery wastewater, this value was 8.80% (w/w), which corresponds to a biodiesel productivity of 23.56 mg L^{-1} d^{-1}. Assemany et al. (2016) evaluated the energy production potential, in terms of lipids and biogas, of microalgae grown in a meat-processing industry wastewater. The studied cultures were able to achieve lipid productivities of 10.0 mg L^{-1} d^{-1}, and methane yields of 2.38 m^3 CH_4 kg^{-1} total volatile solids. At the same time, these cultures have effectively removed NH_4-N and PO_4-P from the wastewater, achieving nitrogen and phosphorus removal efficiencies of 100 and 47.2%, respectively.

Table 2.5. Application of microalgae in nitrogen and phosphorus removal from anaerobically-digested wastewaters and respective removal rates.

Wastewater source	Microorganisms	Nitrogen		Phosphorus		Reference
		C_i (mg N L^{-1})	RR (mg N L^{-1} d^{-1})	C_i (mg P L^{-1})	RR (mg P L^{-1} d^{-1})	
Dairy industry wastewater	Native microalgae	78 TN	6.91	7 TP	0.7	(Wilkie and Mulbry 2002)
Piggery wastewater	*Spirulina* sp.	1209–1481 NH_4-N	145–203	164–620 PO_4-P	17–77	(Olguín et al. 2003)
Dairy industry wastewater	Microalgal consortium	16.3–30.5 NH_3-N	1.04–1.95	1.8–2.6 PO_4-P	0.12–0.17	(Woertz et al. 2009)
Piggery wastewater	*Chlorella* sp.	60 NH_4-N	4.08	18.11 PO_4-P	1.10	(Ledda et al. 2015)
Toilet wastewater	*Chlorella vulgaris/ Chlorella sorokiniana/ Tetradesmus obliquus* (formerly *Scenedesmus obliquus*)	107 NH_4-N	3.09–3.57	73 PO_4-P	1.78–7.54	(Fernandes et al. 2015)
Piggery wastewater	*Chlorella* sp./ *Scenedesmus* sp.	900 NH_4-N	13.3–39.2	n.a.	n.a.	(Nwoba et al. 2016)

C_i—initial concentration (mg L^{-1}); RR—removal rate (mg L^{-1} d^{-1}).

5.3.4 Anaerobically-digested wastewaters

Anaerobically-digested wastewaters typically present large amounts of nitrogen and phosphorus, requiring further treatment before discharge into water courses. The use of microalgae for the remediation of different anaerobically-digested effluents has already been reported in the literature (Table 2.5). For example, Wilkie and Mulbry (2002) have evaluated nutrients recovery in an anaerobically-digested dairy industry effluent using native microalgae, reporting TN and TP removal efficiencies of 62 and 70%, respectively (initial TN and TP concentrations were 78 and 7 mg L^{-1}, respectively). On the other hand, Olguín et al. (2003) and Ledda et al. (2015) have focused on nutrients removal and biomass production in anaerobically-digested piggery wastewaters. In the study performed by Olguín et al. (2003), cultivation of *Spirulina* sp. in outdoor conditions and semi-continuous mode has resulted in biomass productivities between 11.8 and 15.1 g m^{-2} d^{-1}. In the same conditions, NH_4-N removal efficiencies ranged between 84 and 96% and PO_4-P removal efficiencies ranged between 72 and 87% (initial NH_4-N and PO_4-P concentrations were about 1209–1481 and 164–620 mg L^{-1}, respectively). Similarly, Ledda et al. (2015) have grown *Chlorella* sp. in an anaerobically-digested piggery wastewater (with NH_4-N and PO_4-P initial concentrations of 60 and 18.11 mg L^{-1}, respectively), reporting biomass productivities of 0.10 g L^{-1} d^{-1} and nitrogen and phosphorus removal efficiencies of 95 and 85%, respectively. Fernandes et al. (2015) have evaluated the potential of *C. vulgaris*, *C. sorokiniana* and *T. obliquus* in nutrients removal from an anaerobically-digested toilet wastewater (initial NH_4-N and PO_4-P concentrations of 107 and 73 mg L^{-1}, respectively), achieving nitrogen removal efficiencies ranging between 26–30% and phosphorus removal efficiencies ranging between 22–93%.

More recently, Nwoba et al. (2016) have compared microalgal growth in a tubular photobioreactor and in an open pond for treating an anaerobically-digested piggery effluent under outdoor conditions. The results have shown that microalgal growth in the tubular photobioreactor was 2.1 times higher than in the raceway pond. However, no significant differences were observed in terms of NH_4-N removal, since nitrogen removal rates determined in the tubular reactor and in the open pond were 24.6 and 25.9 mg L^{-1} d^{-1}, respectively. Accordingly, the authors have concluded that the studied microalgae (*Chlorella* sp. and *Scenedesmus* sp.) can be effectively applied for nutrients removal and biomass production from an anaerobically-digested piggery wastewater.

5.3.5 Leachate wastewaters

Due to the high NH_4-N concentrations commonly found in leachate wastewaters, and due to its toxicity (Lin et al. 2007), only a few studies have reported microalgal culturing in these effluents. Table 2.6 presents an overview of some studies where microalgae have been applied for nitrogen and phosphorus removal from leachate wastewaters. Taking into account the possible harmful effects of high NH_4-N levels, Lin et al. (2007) have isolated two microalgal strains [*Auxenochlorella pyrenoidosa* (formerly *Chlorella pyrenoidosa*) and *Chlamydomonas snowiae*] from a high ammoniacal leachate environment (Li Keng Landfill, Guangzhou, China) to assess NH_4-N toxicity, and to evaluate the feasibility of NH_4-N tolerant microalgae to treat a landfill leachate. The growth of these microalgae in serial dilutions of landfill leachate has demonstrated that for NH_4-N concentrations higher than 670 mg N L^{-1}, microalgal growth was inhibited. However, for more diluted leachate wastewaters, significant amounts of NH_4-N and PO_4-P were removed by both microalgae (average removal rates ranged between 16.8 and 89.7 mg N L^{-1} d^{-1}, for nitrogen, and between 0.09 and 0.34 mg P L^{-1} d^{-1}, for phosphorus). Paskuliakova et al. (2016) have evaluated if microalgal phycoremediation could be considered a viable biological treatment option for landfill leachates. Accordingly, the authors have grown *Chlamydomonas* sp. and *Scenedesmus* sp. strains (obtained from treated and untreated

Table 2.6. Application of microalgae in nitrogen and phosphorus removal from leachate wastewaters and respective removal rates.

Wastewater source	Microorganisms	Nitrogen		Phosphorus		Reference
		C_i (mg N L^{-1})	RR (mg N L^{-1} d^{-1})	C_i (mg P L^{-1})	RR (mg P L^{-1} d^{-1})	
Landfill leachate	*Auxenochlorella pyrenoidosa* (formerly *Chlorella pyrenoidosa*)/ *Chlamydomonas snowiae*	1345 NH_4-N	16.8– 89.7	5.13 PO_4-P	0.09– 0.34	(Lin et al. 2007)
Landfill leachate	*Auxenochlorella pyrenoidosa* (formerly *Chlorella pyrenoidosa*)	1381 NH_4-N	109.3	3.2 PO_4-P	0.25	(Zhao et al. 2014)
Raw poultry litter leachate	*Arthrospira platensis*/ *Chlorella vulgaris*	279 NH_4-N	n.a.	154 PO_4-P	13.86– 13.94	(Markou et al. 2016)
Landfill leachate	*Chlamydomonas* sp./*Scenedesmus* sp.	88 NH_4-N	3.67	1 PO_4-P	0.07	(Paskuliakova et al. 2016)

C_i—initial concentration (mg L^{-1}); RR—removal rate (mg L^{-1} d^{-1}).

leachate) in different concentrations of a leachate wastewater obtained from a landfill located in Northern Ireland. To reduce energetic costs, cultures were grown at a relatively low temperature (15°C) and light intensity (22 μmol m^{-2} s^{-1}). The results have shown that *Chlamydomonas* sp. strain isolated from untreated leachate achieved the highest NH_4-N removal efficiency (51.7%, for 10% raw leachate). However, when *Chlamydomonas* sp. was grown in 10% raw leachate supplemented with phosphate, NH_4-N removal efficiency increased to 90.7%, indicating that phosphorus can be a limiting factor in landfill leachate remediation by microalgae.

As it was reported for other wastewaters, microalgal culture in landfill leachate has also been used for other purposes, such as CO_2 uptake and biomass and lipid production. In the study performed by Zhao et al. (2014), *A. pyrenoidosa* was cultured in different concentrations of a landfill leachate collected from Laogang landfill (Shanghai, China) to assess its potential in nutrients removal, CO_2 capture, and lipid production. In this study, maximum biomass concentration, 1.58 g L^{-1}, was achieved for 10% raw leachate. In the same conditions, total nitrogen removal efficiency obtained was 90%, and maximum lipid productivity and carbon fixation rate were 24.1 and 65.8 mg L^{-1} d^{-1}, respectively. Markou et al. (2016) have grown *Arthrospira platensis* and *C. vulgaris* in different dilution factor of 10-D10, 15-D15, 20-D20 and 25-D25 of a poultry litter leachate to evaluate growth and lipid productivities achieved by these microorganisms in this type of wastewaters. This study has shown that *A. platensis* was not able to grow in D10 and D15 leachate. However, it was able to grow in D20 and D25 diluted poultry litter leachate, achieving lipid contents of 19.7 and 15.5% (w/w), respectively. On the other hand, *C. vulgaris* was able to grow in all tested dilutions, achieving lipid contents of about 24.2–27.2% (w/w), which were significantly higher than those achieved in control medium (14.1% w/w).

6. Research Needs

Microalgae have shown high potential for biofuel production. However, the integration of microalgal culture with wastewater treatment seems to be mandatory to reduce the production costs (as well as its associated environmental impact in terms of freshwater use), and to be economically competitive with other biofuel technologies. Future research should be focused on the identification of potential effluents (from different sources or different wastewater treatment steps) that present qualitative and quantitative profile of nutrients suitable for microalgal growth. Due to their metabolic activities, microalgal strains have different nutrient requirements. Thus, different species can be selected depending on the composition of the effluent to be treated (Silva et al. 2015). For effluents with high variability in their composition, a consortium of microalgae with microalgae or bacteria should be studied to achieve high nutrient removal efficiencies. The consortium will naturally adapt to the changes in the medium composition.

Another important research topic on microalgal application in WWTPs is the improvement of growth kinetics, and consequently the reduction of hydraulic retention time. This parameter in wastewater treatment is strongly associated with the efficiency and cost of the treatment. Bioreactors should be designed to provide the optimal conditions for microalgal growth, promoting a quick removal of nutrients from the effluent.

7. Conclusion

The search for alternative and renewable energy sources has revealed the potential of photosynthetic microorganisms as a feedstock for biofuels production. However, large scale production of these microorganisms still faces some economical limitations. As a way to reduce biomass production costs, several authors have reported the use of wastewaters as a source of nutrients for microalgal growth. The use of low quality waters for microalgal growth improves both the economic feasibility and environmental sustainability of the overall process, since it reduces the costs associated to nutrients supply and the requirements for freshwater and provides wastewaters' bioremediation. This chapter has demonstrated that microalgae can be effectively cultured in wastewaters from different sources for both removal of nutrients and biomass and biofuels production purposes.

Acknowledgements

This work was financially supported by: Project UID/EQU/00511/2013-LEPABE (Laboratory for Process Engineering, Environment, Biotechnology and Energy—EQU/00511) by FEDER funds through Programa Operacional Competitividade e Internacionalização—COMPETE2020 and by national funds through FCT—Fundação para a Ciência e a Tecnologia; SFRH/BD/88799/2012; V.J.P. Vilar acknowledges the FCT Investigator 2013 Programme (IF/00273/2013). J.C.M. Pires acknowledges the FCT Investigator 2015 Programme (IF/01341/2015).

Keywords: Biofuel; microalgae; nutrient removal; process integration; sustainability; wastewater treatment

References

Adarme-Vega, T.C., D.K.Y. Lim, M. Timmins, F. Vernen, Y. Li and P.M. Schenk. 2012. Microalgal biofactories: a promising approach towards sustainable omega-3 fatty acid production. Microb. Cell Fact. 11: 96–105.

An, J.-Y., S.-J. Sim, J.S. Lee and B.W. Kim. 2003. Hydrocarbon production from secondarily treated piggery wastewater by the green alga *Botryococcus braunii*. J. Appl. Phycol. 15: 185–191.

Assemany, P.P., M.L. Calijuri, M.D. Tango and E.A. Couto. 2016. Energy potential of algal biomass cultivated in a photobioreactor using effluent from a meat processing plant. Algal Res. 17: 53–60.

Azov, Y. 1982. Effect of pH on inorganic carbon uptake in algal cultures. Appl. Environ. Microbiol. 43: 1300–1306.

Barbosa, M.J., M. Albrecht and R.H. Wijffels. 2003. Hydrodynamic stress and lethal events in sparged microalgae cultures. Biotechnol. Bioeng. 83: 112–120.

Benemann, J.R. 1997. CO_2 mitigation with microalgae systems. Energy Convers. Manage. 38: S475–S479.

Bleeke, F., V.M. Rwehumbiza, D. Winckelmann and G. Klock. 2014. Isolation and characterization of new temperature tolerant microalgal strains for biomass production. Energies 7: 7847–7856.

Borowitzka, M.A. 1999. Commercial production of microalgae: ponds, tanks, tubes and fermenters. J. Biotechnol. 70: 313–321.

Brennan, L. and P. Owende. 2010. Biofuels from microalgae—a review of technologies for production, processing, and extractions of biofuels and co-products. Renew. Sust. Energy Rev. 14: 557–577.

Brindley, C., F.G.A. Fernandez and J.M. Fernandez-Sevilla. 2011. Analysis of light regime in continuous light distributions in photobioreactors. Bioresour. Technol. 102: 3138–3148.

Buruiana, D.L., D. Lefter, G.L. Tiron, S. Balta and M. Bordei. 2015. Toxicity of heavy metals on the environment and human health. Int. Multi. Sci. Geoco. 565–571.

Cai, T., S.Y. Park and Y.B. Li. 2013. Nutrient recovery from wastewater streams by microalgae: status and prospects. Renew. Sust. Energy Rev. 19: 360–369.

Caporgno, M.P., A. Taleb, M. Olkiewicz, J. Font, J. Pruvost, J. Legrand et al. 2015. Microalgae cultivation in urban wastewater: nutrient removal and biomass production for biodiesel and methane. Algal Res. 10: 232–239.

Carvalho, A.P., L.A. Meireles and F.X. Malcata. 2006. Microalgal reactors: a review of enclosed system designs and performances. Biotechnol. Progr. 22: 1490–1506.

Cheah, W.Y., P.L. Show, J.-S. Chang, T.C. Ling and J.C. Juan. 2015. Biosequestration of atmospheric CO_2 and flue gas-containing CO_2 by microalgae. Bioresource Technol. 184: 190–201.

Cheirsilp, B. and S. Torpee. 2012. Enhanced growth and lipid production of microalgae under mixotrophic culture condition: effect of light intensity, glucose concentration and fed-batch cultivation. Bioresour. Technol. 110: 510–516.

Chhetri, A.B., M.S. Tango, S.M. Budge, K.C. Watts and M.R. Islam. 2008. Non-edible plant oils as new sources for biodiesel production. Int. J. Mol. Sci. 9: 169–180.

Chinnasamy, S., B. Ramakrishnan, A. Bhatnagar and K.C. Das. 2009. Biomass production potential of a wastewater alga *Chlorella vulgaris* ARC 1 under elevated levels of CO_2 and temperature. Int. J. Mol. Sci. 10: 518–532.

Chinnasamy, S., A. Bhatnagar, R.W. Hunt and K. Das. 2010. Microalgae cultivation in a wastewater dominated by carpet mill effluents for biofuel applications. Bioresour. Technol. 101: 3097–3105.

Chisti, Y. 2007. Biodiesel from microalgae. Biotechnol. Adv. 25: 294–306.

Clark, J.H., R. Luque and A.S. Matharu. 2012. Green chemistry, biofuels, and biorefinery. Annu. Rev. Chem. Biomol. Eng. 3: 183–207.

Covarrubias, S.A., L.E. de-Bashan, M. Moreno and Y. Bashan. 2012. Alginate beads provide a beneficial physical barrier against native microorganisms in wastewater treated with immobilized bacteria and microalgae. Appl. Microbiol. Biotechnol. 93: 2669–2680.

Davis, M. 2011. Water and wastewater engineering: design principles and practice. Mc Graw Hill, New York.

De-Bashan, L.E., A. Trejo, V.A.R. Huss, J.P. Hernandez and Y. Bashan. 2008. *Chlorella sorokiniana* UTEX 2805, a heat and intense, sunlight-tolerant microalga with potential for removing ammonium from wastewater. Bioresour. Technol. 99: 4980–4989.

de Wilt, A., A. Butkovskyi, K. Tuantet, L.H. Leal, T.V. Fernandes, A. Langenhoff et al. 2016. Micropollutant removal in an algal treatment system fed with source separated wastewater streams. J. Hazard Mater 304: 84–92.

Demirbas, A. and M.F. Demirbas. 2011. Importance of algae oil as a source of biodiesel. Energy Convers. Manage. 52: 163–170.

Eggen, R.I.L., J. Hollender, A. Joss, M. Scharer and C. Stamm. 2014. Reducing the discharge of micropollutants in the aquatic environment: the benefits of upgrading wastewater treatment plants. Environ. Sci. Technol. 48: 7683–7689.

Eriksen, N.T. 2008. The technology of microalgal culturing. Biotechnol. Lett. 30: 1525–1536.

Fernandes, T.V., R. Shrestha, Y. Sui, G. Papini, G. Zeeman, L.E.M. Vet et al. 2015. Closing domestic nutrient cycles using microalgae. Environ. Sci. Technol. 49: 12450–12456.

Filippino, K.C., M.R. Mulholland and C.B. Bott. 2015. Phycoremediation strategies for rapid tertiary nutrient removal in a waste stream. Algal Res. 11: 125–133.

Forti, G., A. Furia, P. Bombelli and G. Finazzi. 2003. *In vivo* changes of the oxidation-reduction state of NADP and of the ATP/ADP cellular ratio linked to the photosynthetic activity in *Chlamydomonas reinhardtii*. Plant Physiol. 132: 1464–1474.

Foyer, C.H. and S. Shigeoka. 2011. Understanding oxidative stress and antioxidant functions to enhance photosynthesis. Plant Physiol. 155: 93–100.

Fu, F.L. and Q. Wang. 2011. Removal of heavy metal ions from wastewaters: a review. J. Environ. Manage. 92: 407–418.

Gaffney, J.S. and N.A. Marley. 2009. The impacts of combustion emissions on air quality and climate—from coal to biofuels and beyond. Atmos. Environ. 43: 23–36.

Garcia-Camacho, F., A. Sanchez-Miron, E. Molina-Grima, F. Camacho-Rubio and J.C. Merchuck. 2012. A mechanistic model of photosynthesis in microalgae including photoacclimation dynamics. J. Theor. Biol. 304: 1–15.

Ge, Y.M., J.Z. Liu and G.M. Tian. 2011. Growth characteristics of *Botryococcus braunii* 765 under high CO_2 concentration in photobioreactor. Bioresour. Technol. 102: 130–134.

Giordano, M. and L. Prioretti. 2016. Sulphur and algae: metabolism, ecology and evolution. pp. 185–209. The physiology of microalgae. Springer, Heidelberg. http://link.springer.com/chapter/10.1007%2F978-3-319-24945-2_9

Goldman, J.C. 1973. Carbon dioxide and pH-effect on species succession of algae. Science 182: 306–307.

Gómez-Serrano, C., M. Morales-Amaral, F. Acién, R. Escudero, J. Fernández-Sevilla and E. Molina-Grima. 2015. Utilization of secondary-treated wastewater for the production of freshwater microalgae. Appl. Microbiol. Biotechnol. 99: 6931–6944.

Gonçalves, A.L., M. Simões and J.C.M. Pires. 2014. The effect of light supply on microalgal growth, CO_2 uptake and nutrient removal from wastewater. Energy Convers. Manage. 85: 530–536.

Gonçalves, A.L., J.C.M. Pires and M. Simões. 2016a. Wastewater polishing by consortia of *Chlorella vulgaris* and activated sludge native bacteria. J. Clean. Prod. 133: 348–357.

Gonçalves, A.L., C.M. Rodrigues, J.C.M. Pires and M. Simões. 2016b. The effect of increasing CO_2 concentrations on its capture, biomass production and wastewater bioremediation by microalgae and cyanobacteria. Algal Res. 14: 127–136.

Gouveia, L., S. Graça, C. Sousa, L. Ambrosano, B. Ribeiro, E.P. Botrel et al. 2016. Microalgae biomass production using wastewater: Treatment and costs: Scale-up considerations. Algal Res. 16: 167–176.

Grima, E.M., E.H. Belarbi, F.G.A. Fernandez, A.R. Medina and Y. Chisti. 2003. Recovery of microalgal biomass and metabolites: process options and economics. Biotechnol. Adv. 20: 491–515.

Gui, M.M., K.T. Lee and S. Bhatia. 2008. Feasibility of edible oil vs. non-edible oil vs. waste edible oil as biodiesel feedstock. Energy 33: 1646–1653.

Guiry, M.D. and G.M. Guiry. 2016. AlgaeBase, World-wide electronic publication, National University of Ireland, Galway. Available online at: http://algaebase.org/.

Haines, A., R.S. Kovats, D. Campbell-Lendrum and C. Corvalan. 2006. Harben Lecture - Climate change and human health: impacts, vulnerability, and mitigation. Lancet 367: 2101–2109.

Hansen, P.J. 2002. Effect of high pH on the growth and survival of marine phytoplankton: implications for species succession. Aquat. Microb. Ecol. 28: 279–288.

Henze, M. and Y. Comeau. 2008. Wastewater Characterization. pp. 33–52. Biological wastewater treatment: principles, modelling and design. IWA Publishing, Londres. http://ocw.unesco-ihe.org/pluginfile.php/462/mod_resource/content/1/Urban_Drainage_and_Sewerage/5_Wet_Weather_and_Dry_Weather_Flow_Characterisation/DWF_characterization/Notes/Wastewater%20characterization.pdf

Heredia-Arroyo, T., W. Wei, R. Ruan and B. Hu. 2011. Mixotrophic cultivation of *Chlorella vulgaris* and its potential application for the oil accumulation from non-sugar materials. Biomass Bioenerg. 35: 2245–2253.

Hernandez, J.P., L.E. De-Bashan, D.J. Rodriguez, Y. Rodriguez and Y. Bashan. 2009. Growth promotion of the freshwater microalga *Chlorella vulgaris* by the nitrogen-fixing, plant growth-promoting bacterium *Bacillus pumilus* from and zone soils. Eur. J. Soil Biol. 45: 88–93.

Hernández, D., B. Riaño, M. Coca and M. García-González. 2013. Treatment of agro-industrial wastewater using microalgae-bacteria consortium combined with anaerobic digestion of the produced biomass. Bioresour. Technol. 135: 598–603.

Hirooka, T., Y. Akiyama, N. Tsuji, T. Nakamura, H. Nagase, K. Hirata et al. 2003. Removal of hazardous phenols by microalgae under photoautotrophic conditions. J. Biosci. Bioeng. 95: 200–203.

Ho, S.H., Y.Y. Lai, C.Y. Chiang, C.N.N. Chen and J.S. Chang. 2013. Selection of elite microalgae for biodiesel production in tropical conditions using a standardized platform. Bioresour. Technol. 147: 135–142.

Hu, Q., M. Sommerfeld, E. Jarvis, M. Ghirardi, M. Posewitz, M. Seibert et al. 2008. Microalgal triacylglycerols as feedstocks for biofuel production: perspectives and advances. Plant J. 54: 621–639.

Hu, C.W., L.T. Chuang, P.C. Yu and C.N.N. Chen. 2013. Pigment production by a new thermotolerant microalga *Coelastrella* sp. F50. Food Chem. 138: 2071–2078.

Hulme, M. 2016. 1.5°C and climate research after the Paris Agreement. Nat. Clim. Chang. 6: 222–224.

Iyer, G., Y. Gupte, P. Vaval and V. Nagle. 2015. Uptake of potassium by algae and potential use as biofertilizer. Ind. J. Plant Physiol. 20: 285–288.

Jacob-Lopes, E., C.H.G. Scoparo, L.M.C.F. Lacerda and T.T. Franco. 2009. Effect of light cycles (night/day) on CO_2 fixation and biomass production by microalgae in photobioreactors. Chem. Eng. Process 48: 306–310.

Janssen, M., P. Slenders, J. Tramper, L.R. Mur and R.H. Wijffels. 2001. Photosynthetic efficiency of *Dunaliella tertiolecta* under short light/dark cycles. Enzyme Microb. Technol. 29: 298–305.

Janssen, M., J. Tramper, L.R. Mur and R.H. Wijffels. 2003. Enclosed outdoor photobioreactors: light regime, photosynthetic efficiency, scale-up, and future prospects. Biotechnol. Bioeng. 81: 193–210.

Ji, M.K., A.N. Kabra, J. Choi, J.H. Hwang, J.R. Kim, R.A.I. Abou-Shanab et al. 2014. Biodegradation of bisphenol A by the freshwater microalgae *Chlamydomonas mexicana* and *Chlorella vulgaris*. Ecol. Eng. 73: 260–269.

Juneja, A., R.M. Ceballos and G.S. Murthy. 2013. Effects of environmental factors and nutrient availability on the biochemical composition of algae for biofuels production: a review. Energies 6: 4607–4638.

Keren, N., M.J. Kidd, J.E. Penner-Hahn and H.B. Pakrasi. 2002. A light-dependent mechanism for massive accumulation of manganese in the photosynthetic bacterium *Synechocystis* sp. PCC 6803. Biochemistry 41: 15085–15092.

Khan, M.N. and F. Mohammad. 2014. Eutrophication: challenges and solutions. pp. Eutrophication: causes, consequences and control. Springer Science, Dordrecht.

Kjeldsen, P., M.A. Barlaz, A.P. Rooker, A. Baun, A. Ledin and T.H. Christensen. 2002. Present and long-term composition of MSW landfill leachate: a review. Crit. Rev. Environ. Sci. Technol. 32: 297–336.

Kong, Q.-x., L. Li, B. Martinez, P. Chen and R. Ruan. 2009. Culture of microalgae *Chlamydomonas reinhardtii* in wastewater for biomass feedstock production. Appl. Biochem. Biotechnol. 160: 9–18.

Kumar, A., S. Ergas, X. Yuan, A. Sahu, Q.O. Zhang, J. Dewulf et al. 2010. Enhanced CO_2 fixation and biofuel production via microalgae: recent developments and future directions. Trends Biotechnol. 28: 371–380.

Ledda, C., A. Idà, D. Allemand, P. Mariani and F. Adani. 2015. Production of wild *Chlorella* sp. cultivated in digested and membrane-pretreated swine manure derived from a full-scale operation plant. Algal Res. 12: 68–73.

Lewis, S.L. 2016. The Paris Agreement has solved a troubling problem. Nature 532: 283–283.

Li, X., H.Y. Hu, K. Gan and Y.X. Sun. 2010. Effects of different nitrogen and phosphorus concentrations on the growth, nutrient uptake, and lipid accumulation of a freshwater microalga *Scenedesmus* sp. Bioresour. Technol. 101: 5494–5500.

Li, Y.R., W.T. Tsai, Y.C. Hsu, M.Z. Xie and J.J. Chen. 2014. Comparison of autotrophic and mixotrophic cultivation of green microalgal for biodiesel production. Energy Procedia 52: 371–376.

Lim, S.-L., W.-L. Chu and S.-M. Phang. 2010. Use of *Chlorella vulgaris* for bioremediation of textile wastewater. Bioresour. Technol. 101: 7314–7322.

Lin, L., G. Chan, B. Jiang and C. Lan. 2007. Use of ammoniacal nitrogen tolerant microalgae in landfill leachate treatment. Waste Manage. 27: 1376–1382.

Liu, J., B. Danneels, P. Vanormelingen and W. Vyverman. 2016. Nutrient removal from horticultural wastewater by benthic filamentous algae *Klebsormidium* sp., *Stigeoclonium* spp. and their communities: From laboratory flask to outdoor Algal Turf Scrubber (ATS). Water Res. 92: 61–68.

Maestrini, S.Y., J.M. Robert, J.W. Leftley and Y. Collos. 1986. Ammonium thresholds for simultaneous uptake of ammonium and nitrate by oyster-pond algae. J. Exp. Mar. Biol. Ecol. 102: 75–98.

Markou, G., D. Iconomou and K. Muylaert. 2016. Applying raw poultry litter leachate for the cultivation of *Arthrospira platensis* and *Chlorella vulgaris*. Algal Res. 13: 79–84.

Mehta, S.K. and J.P. Gaur. 2005. Use of algae for removing heavy metal ions from wastewater: progress and prospects. Crit. Rev. Biotechnol. 25: 113–152.

Menetrez, M.Y. 2012. An overview of algae biofuel production and potential environmental impact. Environ. Sci. Technol. 46: 7073–7085.

Mennaa, F.Z., Z. Arbib and J.A. Perales. 2015. Urban wastewater treatment by seven species of microalgae and an algal bloom: Biomass production, N and P removal kinetics and harvestability. Water Res. 83: 42–51.

Monks, P.S., C. Granier, S. Fuzzi, A. Stohl, M.L. Williams, H. Akimoto et al. 2009. Atmospheric composition change—global and regional air quality. Atmos. Environ. 43: 5268–5350.

Mutanda, T., D. Ramesh, S. Karthikeyan, S. Kumari, A. Anandraj and F. Bux. 2011. Bioprospecting for hyper-lipid producing microalgal strains for sustainable biofuel production. Bioresour. Technol. 102: 57–70.

Nayak, M., A. Karemore and R. Sen. 2016. Performance evaluation of microalgae for concomitant wastewater bioremediation, CO_2 biofixation and lipid biosynthesis for biodiesel application. Algal Res. 16: 216–223.

Norsker, N.H., M.J. Barbosa, M.H. Vermue and R.H. Wijffels. 2011. Microalgal production—a close look at the economics. Biotechnol. Adv. 29: 24–27.

Nwoba, E.G., J.M. Ayre, N.R. Moheimani, B.E. Ubi and J.C. Ogbonna. 2016. Growth comparison of microalgae in tubular photobioreactor and open pond for treating anaerobic digestion piggery effluent. Algal Res. 17: 268–276.

Olguín, E.J., S. Galicia, G. Mercado and T. Pérez. 2003. Annual productivity of *Spirulina* (*Arthrospira*) and nutrient removal in a pig wastewater recycling process under tropical conditions. J. Appl. Phycol. 15: 249–257.

Orpez, R., M.E. Martinez, G. Hodaifa, F. El Yousfi, N. Jbari and S. Sanchez. 2009. Growth of the microalga *Botryococcus braunii* in secondarily treated sewage. Desalination 246: 625–630.

Paskuliakova, A., S. Tonry and N. Touzet. 2016. Phycoremediation of landfill leachate with chlorophytes: phosphate a limiting factor on ammonia nitrogen removal. Water Res. 99: 180–187.

Perez-Garcia, O., F.M.E. Escalante, L.E. de-Bashan and Y. Bashan. 2011. Heterotrophic cultures of microalgae: metabolism and potential products. Water Res. 45: 11–36.

Pienkos, P.T. and A. Darzins. 2009. The promise and challenges of microalgal-derived biofuels. Biofuel. Bioprod. Biorefin. 3: 431–440.

Pires, J.C.M., M.C.M. Alvim-Ferraz, F.G. Martins and M. Simões. 2012. Carbon dioxide capture from flue gases using microalgae: engineering aspects and biorefinery concept. Renew. Sust. Energy Rev. 16: 3043–3053.

Pires, J., M. Alvim-Ferraz, F. Martins and M. Simões. 2013. Wastewater treatment to enhance the economic viability of microalgae culture. Environ. Sci. Pollut. Res. 20: 5096–5105.

Pires, J.C.M., A.L. Gonçalves, F.G. Martins, M.C.M. Alvim-Ferraz and M. Simões. 2014. Effect of light supply on CO_2 capture from atmosphere by *Chlorella vulgaris* and *Pseudokirchneriella subcapitata*. Mitig. Adapt. Strateg. Glob. Chang. 19: 1109–1117.

Pittman, J.K., A.P. Dean and O. Osundeko. 2011. The potential of sustainable algal biofuel production using wastewater resources. Bioresour. Technol. 102: 17–25.

Posten, C. 2009. Design principles of photo-bioreactors for cultivation of microalgae. Eng. Life Sci. 9: 165–177.

Powell, N., A.N. Shilton, S. Pratt and Y. Chisti. 2008. Factors influencing luxury uptake of phosphorus by microalgae in waste stabilization ponds. Environ. Sci. Technol. 42: 5958–5962.

Powell, N., A. Shilton, Y. Chisti and S. Pratt. 2009. Towards a luxury uptake process via microalgae—defining the polyphosphate dynamics. Water Res. 43: 4207–4213.

Pruder, G.D. and E.T. Bolton. 1979. Role of CO_2 enrichment of aerating gas in the growth of an estuarine diatom. Aquaculture 17: 1–15.

Pulz, O. and W. Gross. 2004. Valuable products from biotechnology of microalgae. Appl. Microbiol. Biotechnol. 65: 635–648.

Queiroz, M.I., E.J. Lopes, L.Q. Zepka, R.G. Bastos and R. Goldbeck. 2007. The kinetics of the removal of nitrogen and organic matter from parboiled rice effluent by cyanobacteria in a stirred batch reactor. Bioresour. Technol. 98: 2163–2169.

Raja, R., S. Hemaiswarya, N.A. Kumar, S. Sridhar and R. Rengasamy. 2008. A perspective on the biotechnological potential of microalgae. Crit. Rev. Microbiol. 34: 77–88.

Rawat, I., R.R. Kumar, T. Mutanda and F. Bux. 2011. Dual role of microalgae: Phycoremediation of domestic wastewater and biomass production for sustainable biofuels production. Appl. Energy 88: 3411–3424.

Redfield, A.C. 1958. The biological control of chemical factors in the environment. Am. Scientist 46: 205–221.

Richardson, J.W., M.D. Johnson, X.Z. Zhang, P. Zemke, W. Chen and Q. Hu. 2014. A financial assessment of two alternative cultivation systems and their contributions to algae biofuel economic viability. Algal Res. 4: 96–104.

Riding, R. 2009. An atmospheric stimulus for cyanobacterial-bioinduced calcification ca. 350 million years ago? Palaios 24: 685–696.

Riebesell, U. and D. Wolf-Gladrow. 2002. Supply and uptake of inorganic nutrients. pp. 109–140. Phytoplankton productivity: carbon assimilation in marine and freshwater ecosystems. Blackwell Publishing Ltd., Oxford, UK. http://onlinelibrary.wiley.com/doi/10.1002/9780470995204.ch5/summary

Rubio, F.C., F.G. Camacho, J.M.F. Sevilla, Y. Chisti and E.M. Grima. 2003. A mechanistic model of photosynthesis in microalgae. Biotechnol. Bioeng. 81: 459–473.

Ruiz, J., Z. Arbib, P. Álvarez-Díaz, C. Garrido-Pérez, J. Barragán and J. Perales. 2013. Photobiotreatment model (PhBT): a kinetic model for microalgae biomass growth and nutrient removal in wastewater. Environ. Technol. 34: 979–991.

Sayre, R. 2010. Microalgae: the potential for carbon capture. Bioscience 60: 722–727.

Sforza, E., D. Simionato, G.M. Giacometti, A. Bertucco and T. Morosinotto. 2012. Adjusted light and dark cycles can optimize photosynthetic efficiency in algae growing in photobioreactors. PlosOne 7.

Shi, J., B. Podola and M. Melkonian. 2007. Removal of nitrogen and phosphorus from wastewater using microalgae immobilized on twin layers: an experimental study. J. Appl. Phycol. 19: 417–423.

Silva, N.F.P., A.L. Gonçalves, F.C. Moreira, T.F.C.V. Silva, F.G. Martins, M.C.M. Alvim-Ferraz et al. 2015. Towards sustainable microalgal biomass production by phycoremediation of a synthetic wastewater: a kinetic study. Algal Res. 11: 350–358.

Singh, G. and P.B. Thomas. 2012. Nutrient removal from membrane bioreactor permeate using microalgae and in a microalgae membrane photoreactor. Bioresour. Technol. 117: 80–85.

Singh, J. and S. Cu. 2010. Commercialization potential of microalgae for biofuels production. Renew. Sust. Energy Rev. 14: 2596–2610.

Slade, R. and A. Bauen. 2013. Micro-algae cultivation for biofuels: cost, energy balance, environmental impacts and future prospects. Biomass Bioenerg. 53: 29–38.

Spellman, F.R. 2013. Handbook of water and wastewater treatment plant operations. CRC Press, Boca Raton, Florida, USA.

Spolaore, P., C. Joannis-Cassan, E. Duran and A. Isambert. 2006. Commercial applications of microalgae. J. Biosci. Bioeng. 101: 87–96.

Su, Y., A. Mennerich and B. Urban. 2012. Synergistic cooperation between wastewater-born algae and activated sludge for wastewater treatment: influence of algae and sludge inoculation ratios. Bioresour. Technol. 105: 67–73.

Sydney, E.B., A.C. Novak, J.C. Carvalho and C.R. Soccol. 2014. Respirometric balance and carbon fixation of industrially important algae. pp. 67–84. *In*: A. Pandey, D.-J. Lee, Y. Chisti, C.R. Soccol (eds.). Biofuels from Algae. Elsevier, Burlington, Massachusetts, USA.

Tchobanoglous, G., F.L. Burton and H.D. Stensel. 2003. Wastewater engineering: treatment and reuse. McGraw Hill, Boston.

Valderrama, L.T., C.M. Del Campo, C.M. Rodriguez, L.E. de-Bashan and Y. Bashan. 2002. Treatment of recalcitrant wastewater from ethanol and citric acid production using the microalga *Chlorella vulgaris* and the macrophyte *Lemna minuscula*. Water Res. 36: 4185–4192.

Van Den Hende, S., E. Carré, E. Cocaud, V. Beelen, N. Boon and H. Vervaeren. 2014. Treatment of industrial wastewaters by microalgal bacterial flocs in sequencing batch reactors. Bioresour. Technol. 161: 245–254.

Vasudevan, V., R.W. Stratton, M.N. Pearlson, G.R. Jersey, A.G. Beyene, J.C. Weissman et al. 2012. Environmental performance of algal biofuel technology options. Environ. Sci. Technol. 46: 2451–2459.

Volesky, B. and Z.R. Holan. 1995. Biosorption of heavy-metals. Biotechnol. Progr. 11: 235–250.

Vonshak, A. and G. Torzillo. 2004. Environmental stress physiology. pp. 57–82. *In*: A. Richmond (ed.). Handbook of Microalgal Culture. Blackwell, Oxford, UK.

Wahidin, S., A. Idris and S.R.M. Shaleh. 2013. The influence of light intensity and photoperiod on the growth and lipid content of microalgae *Nannochloropsis* sp. Bioresour. Technol. 129: 7–11.

Wan, M.X., P. Liu, J.L. Xia, J.N. Rosenberg, G.A. Oyler, M.J. Betenbaugh et al. 2011. The effect of mixotrophy on microalgal growth, lipid content, and expression levels of three pathway genes in *Chlorella sorokiniana*. Appl. Microbiol. Biotechnol. 91: 835–844.

Wang, B., Y.Q. Li, N. Wu and C.Q. Lan. 2008. CO_2 bio-mitigation using microalgae. Appl. Microbiol. Biotechnol. 79: 707–718.

Wang, L., M. Min, Y. Li, P. Chen, Y. Chen, Y. Liu et al. 2010. Cultivation of green algae *Chlorella* sp. in different wastewaters from municipal wastewater treatment plant. Appl. Biochem. Biotechnol. 162: 1174–1186.

Whitton, R., A. Le Mével, M. Pidou, F. Ometto, R. Villa and B. Jefferson. 2016. Influence of microalgal N and P composition on wastewater nutrient remediation. Water Res. 91: 371–378.

Wijffels, R.H. and M.J. Barbosa. 2010. An outlook on microalgal biofuels. Science 329: 796–799.

Wilkie, A.C. and W.W. Mulbry. 2002. Recovery of dairy manure nutrients by benthic freshwater algae. Bioresour. Technol. 84: 81–91.

Williams, P.J.L. and L.M.L. Laurens. 2010. Microalgae as biodiesel & biomass feedstocks: review & analysis of the biochemistry, energetics & economics. Energy Environ. Sci. 3: 554–590.

Woertz, I., A. Feffer, T. Lundquist and Y. Nelson. 2009. Algae grown on dairy and municipal wastewater for simultaneous nutrient removal and lipid production for biofuel feedstock. J. Environ. Eng. 135: 1115–1122.

Wu, L.F., P.C. Chen, A.P. Huang and C.M. Lee. 2012. The feasibility of biodiesel production by microalgae using industrial wastewater. Bioresour. Technol. 113: 14–18.

Yang, Y. and K.S. Gao. 2003. Effects of CO_2 concentrations on the freshwater microalgae, *Chlamydomonas reinhardtii, Chlorella pyrenoidosa* and *Scenedesmus obliquus* (Chlorophyta). J. Appl. Phycol. 15: 379–389.

Zavarzin, G.A. 2005. Recent microbiology and precambrian paleontology. pp. 201–216. *In*: R.B. Hoover, A.Yu. Rozanov and R. Paepe (eds.). Perspectives in Astrobiology. IOS Press.

Zhao, X., Y. Zhou, S. Huang, D. Qiu, L. Schideman, X. Chai et al. 2014. Characterization of microalgae-bacteria consortium cultured in landfill leachate for carbon fixation and lipid production. Bioresour. Technol. 156: 322–328.

Zhu, L., Z. Wang, Q. Shu, J. Takala, E. Hiltunen, P. Feng et al. 2013. Nutrient removal and biodiesel production by integration of freshwater algae cultivation with piggery wastewater treatment. Water Res. 47: 4294–4302.

CHAPTER 3

Biodiesel Production from Microalgae

Sarmidi Amin[1,]* and *Kurniadhi Prabandono*[2]

1. Introduction

In 1900, Rudolf Diesel demonstrated a working diesel engine using peanut oil as fuel at the World Exhibition in Paris (Agarwal 2007, Wikipedia biodiesel). This is the first time people using vegetable based material as a fuel. Then it is called biodiesel. There are many raw materials biodiesel typically uses, including restaurant waste oil, animal fat, and seed oil derived from plants. Using food supply seed oils as sources of biodiesel leads to a higher price of seeds, which causes the cost of biodiesel to become increasingly expensive (Campbell 2008). The potential raw material from plants, such as crude palm oil (CPO) pond, CPO off grade, palm stearin, and *Jatropha curcas* L. are other potential raw materials used as sources of biodiesel. Other food supplies which also have the potential to be used for it are avocado, olive, coconut, peanut, and soybean (Amin 2007). Other raw materials are cottonseed oil, castor oil, and a few less common oils, such as babasu, crude raisin seed oil, industrial tallow, and fish oil (Knothe 2005).

Biodiesel can be also produced from microalgae. Microalgae are not a main source of food; therefore this raw material is relatively cheaper than sources of food. The development of biofuel as substitute petroleum

[1] Agency for Assessment and Application Technology-BPPT (Retired). Home: Komplek Inkoppol Jl. Merak IV No. 27. Jakasampurna Bekasi Barat, Indonesia 17137.
[2] Ministry for Marine Affair and Fisheries-Republic Indonesia, Indonesian Fish Quarantine Center (FQIA). Mina Bahari Building 2, 6th Floor, Medan Merdeka Timur Street no. 16, Gambir, Center of Jakarta, Indonesia 10110.
  Email: prabandono_k@yahoo.co.id
* Corresponding author: sarmidiamin@gmail.com

diesel is receiving great attention among researchers and policy makers for its numerous advantages, such as renewability, biodegradability, and lower gaseous emission profile (Shah et al. 2013). The oils (triglycerides) are converted by transesterification using alcohol and catalyst to yield glycerol and the fatty acid alkyl ester or fatty acid methyl ester (FAME) (Wahlen et al. 2011).

Growing algae as a source of protein on a large scale in open ponds was first conceived by German scientists during World War II. The first attempt in USA to translate the biological requirements for algal growth into engineering specifications for a large scale plant was made at the Stanford Research Institute during 1948–1950 (Demirbas 2010).

Mass culture of microalgae really began to be a focus of research after 1948 at Stanford (USA), Essen (Germany), and Tokyo. In 1952, an Algae Mass Culture Symposium was held at Stanford University, California. One important outcome of this symposium was the publication of "Algae Culture from Laboratory to Pilot Plant", edited by J.S. Burlew (Borowitzka 2013).

The concept of using algae as a fuel was first proposed by Meier in 1955 for production of methane gas from the carbohydrate fractional cells. This idea was further developed by Oswald and Golueke in 1960, who introduced conceptual technoeconomic engineering analysis of digesting microalgal biomass grown in large raceway ponds to produce methane gas (Demirbas and Demirbas 2010). Algae have been suggested as good candidates for fuel production due to their higher photosynthetic efficiency, higher biomass production, and faster growth as compared to other energy crops (Miao and Wu 2004). The productivity rates of microalgae are higher than most other plants-up to 12.5 kg/m^2/yr (Shay 1993).

According to Shay (1993) and Minowa et al. (1995), microalgae usually have a higher photosynthetic efficiency than other biomasses, such as trees. This biomass is a highly promising resource, according to Hall in Sawayama et al. (1999) and Minowa et al. (1995).

Microalgae are unicellular photosynthetic micro-organisms, living in saline or freshwater environments that convert sunlight, water, and carbon dioxide to algal biomass (Ozkurt 2009). There are two main populations of algae: filamentous and phytoplankton algae. The algal organisms are photosynthetic macroalgae, or microalgae growing in aquatic environments.

The Aquatic Species Program (ASP) considered three main options only for fuel production, i.e., methane gas, ethanol, and biodiesel (Sheechan et al. 1998). The fourth option is the direct combustion of the algal biomass for production of steam or electricity, but the ASP did not focus much attention on that.

2. Cultivation of Microalgae

Microalgae cultivation using sunlight energy can be carried out in open ponds, or closed photobioreactors, based on tubular flat plates, or other designs. There are three distinct algae production mechanisms, including photoautotrophic, heterotrophic, and mixotrophic production. Photoautotrophic production is autotrophic photosynthesis, heterotrophic production requires organic substance (e.g., glucose) to stimulate growth, while some algae strains can combine autotrophic photosynthesis and heterotrophic assimilation or organic compound, resulting in mixotrophic process (Brennan and Owende 2010).

The production pathway of lipid, as shown in Fig. 3.1 includes cultivation, harvesting, drying, cell disruption, and extraction of algal oil or lipid separation (Beal et al. 2013). The selective microalgae are cultivated in a photobiorector, or in outdoor culture. After harvesting, the algal are concentrated by using a centrifugal separator or other methods such as gravity sedimentation, filtration, microscreening, ultra-filtration, flocculation, and coagulation. Beal et al. (2013) reported that the concentrated algae were exposed to electromechanical pulsing in a process designed to lyse the cell. After lysing, the algae were again left in a dark container at room temperature (~ 24°C) overnight (for about 20 h). Other researchers (Alcaine 2010) continued to dry the algae after the concentration process. Several methods have been employed to dry the algae, including spray drying, freeze drying, or sun drying. After drying, it follows the cell disruption of microalgae. Several methods can be used such as bead mill, homogenizer, freezing, sonication, and then alga oil can be extracted using chemical or mechanical methods.

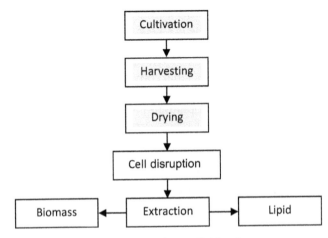

Figure 3.1. The production lipid pathway.

2.1 Open Ponds

Algae can be cultured in open-ponds (such as raceway ponds and lakes) or photobioreactors. Raceway ponds and lakes are less expensive, but they are highly vulnerable to contamination by other microorganisms, such as other algal species or bacteria. Besides that, open systems also do not offer control over temperature and lighting. The growing season is largely dependent on location and, aside from tropical areas, is limited to the warmer months.

The simple open algae cultivation systems are shallow, with a size range between a few m^2 to 250 ha non-stirred ponds (Bioenergynet cultivation 2013), or use paddle wheels for circulation of the water, and are no more than 30 cm deep (Janssen 2002). The necessary carbon for their growth is received through the atmosphere. The CO_2 transfer from the air to the water through naturally-occurring dissolution phenomena constrains the significant growth of algae, resulting in a low crop yield. Another disadvantage of this system is the slow transfer of nutrients across the whole mass of the crop (Bioenergynet cultivation 2013).

2.2 Closed Bioreactors

A photobioreactor (PBR) is a closed reactor which incorporates a light source and provides all nutrients, including CO_2. A PBR can operate in "batch mode", which involves restocking the reactor after each harvest, but it is also possible to grow and harvest continuously. Continuous operation requires precise control of all elements to prevent immediate collapse. The grower provides sterilized water, nutrients, air, and carbon dioxide at the precise rates. The advantage of this reactor is its ability to operate for long periods (Wikipedia algaculture). There are two types of close system, i.e., flat panel photobioreactor and tubular photobioreactor with air-lift column.

One of the major issues in the cultivation of microalgae is light limitation. The effective photosynthetic zone to the volume is within 5 cm of the surface of a pond. These small volumes allow light to penetrate better, yet it may lead to biofouling (the attachment of organisms to a surface in contact with water for a period of time), and the cost of pumping the algae around through the small volumes increases (Austin biomass magazine).

2.3 Harvesting

The most common microalgae harvesting methods include gravity sedimentation, flocculation, centrifugation, filtration and micro-screening, flotation, and electrophoresis techniques. The selection of harvesting technique is dependent on the properties of the microalgae (Prabandono and Amin 2015a). Microalgae harvesting can generally be divided into a two-step process, including (Brenann and Owende 2010):

- Bulk harvesting used to separate microalgal biomass from bulk suspension.
- Thickening concentrated in the slurry with filtration and centrifugation.

Gravity sedimentation is commonly applied separate microalgae from water. Flocculation is frequently used to increase the efficiency of gravity sedimentation, but addition of flocculation is currently not a method of choice for cheap and sustainable production (Schenk et al. 2008). Auto-flocculation also occurs as a result of interrupting the CO_2 supply to an algal system (Demirbas 2010). Filtration is often applied at a laboratory scale, but in application on a large scale it has problems, such as membrane clogging, the formation of compressible filter cake, and high maintenance costs.

Centrifugation is a preferred method, especially for producing extended shelf-life concentrates for aquaculture; however, this method is time-consuming and costly. Centrifugation is a very useful secondary harvesting method to concentrate initial slurry (10–20 g/L) to an algal paste (100–200 g/L), and could possibly be used in combination with oil extraction (Schenk et al. 2008). Flotation is a gravity separation process in which air or gas bubbles are attached to solid particles, which then carry them to the liquid surface. Zhang et al. (2010) developed an efficient technology for harvesting algal biomass using membrane filtration.

2.4 Oil Extraction

After separation from the culture medium, algal biomass must be quickly processed, because it can spoil in only a few hours in a hot climate (Mata et al. 2010). According to Liu et al. (2011), after harvesting, chemicals in biomass may be subjected to degradation induced by the process itself and also by internal enzymes in algal cells. For example, lipase contained in the cells can rapidly hydrolyze cellular lipids into free fatty acids, which are not suitable for biodiesel production.

Oil extraction from dried biomass can be performed in two steps, i.e., mechanical crushing followed by hexane solvent extraction. Oil extraction from algal cells can also be facilitated by osmotic shock or ultrasonic treatment to break the cells. For cell disruption, besides the mechanical crushing, there are cell homogenizer bead mills, ultrasound, autoclave and spray drying, and non-mechanical crushing that uses freezing, organic solvents, osmotic shock, and acid, base and enzyme reactions (Mata et al. 2010). Lee et al. (2010) investigated several methods for effective lipid extraction from microalgae, including autoclaving, bead beating, microwaves, sonication, and 10% NaCl solution, to identify the most effective cell disruption method. Recent research of bio-oil production by pyrolysis of biomass and liquefaction (Huang et al. 2010, Miao and Wu 2004) has received much interest. Other oil extraction methods are supercritical fluid extraction, enzymatic extraction, osmotic shock, and ultrasonic assisted.

2.4.1 Liquefaction

Microalgal cell precipitates derived from centrifugation, which have a high moisture content, are thus good raw materials for liquefaction (FAO oil production). Direct hydrothermal liquefaction in sub-critical water condition is a technology that can be employed to convert wet biomass to liquid fuel (Patil et al. 2008).

The separation scheme is presented in Fig. 3.2 (Minowa et al. 1995, Yang et al. 2004, Murakami et al. 1990, Itoh et al. 1994, Amin 2009, Prabandono and Amin 2015b). The liquefaction is performed in an aqueous solution of alkali or alkaline earth salt at about 300°C and 10 MPa without a reducing gas such as hydrogen and/or carbon monoxide (Minowa et al. 1995).

Liquefaction can be performed by using a stainless steel autoclave with mechanical mixing. The autoclave is charged with algal cell, following which nitrogen is introduced to purge the residual air. The reaction is initiated by heating the autoclave to a fixed temperature and by elevated nitrogen pressure. The temperature is maintained constant for a 5–6 min period, following which it is cooled with the use of an electric fan.

The reaction is extracted with dichloromethane in order to separate the oil fraction. Then the oil and residual dichloromethane is filtered and

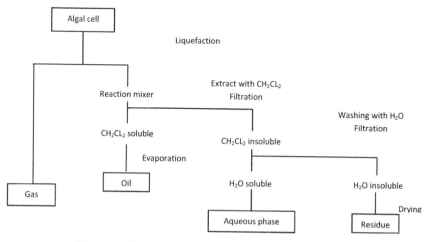

Figure 3.2. Separation scheme for liquefies microalgal cells.

evaporated at 35°C under reduced pressure, yielding a dark brown viscous material (hereafter referred to as the oil). The aqueous phase resulting after dichloromethane extraction (insoluble fraction) is washed with water and filtered from the dichloromethane insoluble (Minowa et al. 1995, FAO oil production).

2.4.2 Fast pyrolysis

Pyrolysis is the conversion of biomass to biofuel, charcoal, and gaseous fraction by heating the biomass in the absence of air to around 500°C (McKendry 2003, Miao et al. 2004), or by heating in the presence of a catalyst (Agarwal 2007) at a high heating rate (10^3–10^4 K/s) and with short gas residence time to crack into short chain molecules and then being cooled to a liquid rapidly (Qi et al. 2007, Bridgwater et al. 1999, Huang et al. 2010, Jain and Sirisha 2015). Previous studies were using slow pyrolysis processes; they were performing at a low heating rate and long residence time. The longer residence time can cause secondary cracking of the primary products, thereby reducing yield and adversely affecting the biofuel properties. In addition, a low heating rate and long residence time may increase the energy input. In recent years, fast pyrolysis processes for biomass have attracted a great deal of attention for maximizing liquid yields, and many researches have been performed (Miao et al. 2004, Qi et al. 2007).

The advantage of fast pyrolysis is that it can directly produce a liquid fuel (Bridgwater and Peacocke 2000), as well as biogas. The products of the fast pyrolysis are oil and gas with a yield of approximately 70% (Huang et al. 2010), or high liquid yield as much as 70 to 80% (Qi et al. 2007). If flash pyrolysis is used, the conversion of biomass to bio-crude with an efficiency of up to 80% is enabled. A conceptual fluidized bed fast pyrolysis system is shown in Fig. 3.3 (Bridgwater and Peacocke 2000). Since microalgae usually have high moisture content, a drying process requires a lot of heating energy (Yang et al. 2004). Algae are subjected to pyrolysis in the fluid bed reactor. The result of the reaction then flows to a cyclone, and it is separated into char, biofuel, and gas. The resultant gas can be used for drying the raw material, or for heating for the process, or as a source of biomethane.

3. Lipid Composition

The compositions of various microalgae are shown in Table 3.1. Table 3.2 shows the lipid content and fatty acids (FA) profile of microalgae. Lipid can be processed to biodiesel, carbohydrates to be ethanol and H_2, and proteins as raw material of biofertilizer. Lipid is the general name for plants and animal products that are structurally ester of higher FA. The FA are any variety of mono basic acid such as palmatic (C16:0), stearic

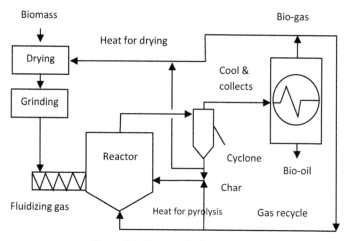

Figure 3.3. Fast pyrolysis principles.

Table 3.1. Composition of selected microalgae (% dry matter basis).

Strain	Protein	Carbohydrates	Lipid
Tetradesmus obliquus	50–55	10–15	12–14
Scenedesmus communis	40	12	1.9
Tetradesmus dimorphus	8–18	21–52	16–40
Chlamydomonas reinhardtii	48	17	21
Chlorella vulgaris	51–58	12–17	14–22
Auxenochlorella pyrenoidosa	57	26	2
Spirogyra sp.	6–20	33–64	11–21
Dunaliella bioculata	49	4	8
Dunaliella salina	57	32	6
Euglena gracilis	39–61	14–18	14–20
Prymnesium parvum	28–45	25–33	22–38
Tetraselmis maculate	52	15	3
Porphyridium purpureum	28–39	40–57	9–14
Arthrospira platensis	46–63	8–14	4–9
Arthrospira maxima	60–71	13–16	6–7
Synechococcus sp.	63	15	11
Anabaena cylindrical	43–56	25–30	4–7

Source: Adapted from Raja et al. (2013)

Table 3.2. Total lipid content and FA profile of microalgae.

Species		Fresh water algae					
		Synechocystis pevalekii	*Nostoc ellipsosporum*	*Arthrospira platensis*	*Spirogyra orientalis*	*Chlorococcum infusionum*	*Rhizoclonium fontinale*
Total Lipid (%)		9 ± 2	7 ± 1.5	8.5 ± 2	21 ± 2.5	11.34 ± 1	7.5 ± 1.4
Fatty Acid (FA)							
Saturated FA	C16:0	34.2	22.05	21.1	34.08	25.62	29.36
Monounsaturated FA	C18:1	29.8	17.85	11.27	6.4	15.66	22.13
Polyunsaturated FA	C18:2	14.9	2.5	0.24	6.7	7.5	–

Species		Brackish water algae						
		Limnoraphis birgei (formerly *Lyngbya birgei*)	*Rhizoclonium riparium*	*Pithophora roettleri* (formerly *Pithophora cleveana*)	*Cladophora crystallina*	*Chaetomorpha gracilis*	*Ulva intestinalis* (formerly *Enteromorpha intestinalis*)	*Polysiphonia mollis*
Total Lipid (%)		12 ± 2.8	8.65 ± 1	19 ± 2	23 ± 1.8	16 ± 0.5	12 ± 1	7.8 ± 1.6
Fatty Acid (FA)								
Saturated FA	C16:0	59.6	37.4	37.0	39.1	52.4	22.57	44.7
Monounsaturated FA	C18:1	13.1	17.3	34.3	31.8	15.9	18.49	10.3
Polyunsaturated FA	C18:2	9.1	17.2	–	11.6	3.6	14.83	–

Species		Marine water algae							
		Leptolyngbya pseudovalderiana (formerly *Phormidium valderianum*)	*Leptolyngbya tenuis* (formerly *Phormidium tenue*)	*Rhizoclonium africanum*	*Ulva lactuca*	*Eolimna minima* (formerly *Navicula minima*)	*Catenella caespitosa* (formerly *Catenella repens*)	*Geledium pasillum*	*Ceramium manorensis*
Total Lipid (%)		7.8 ± 2.8	8.01 ± 2.2	7.2 ± 2.7	11 ± 1	16.23 ± 0.59	8 ± 1.5	9.7 ± 2.8	8 ± 1.9
Fatty Acid (FA)									
Saturated FA	C16:0	35.8	32.9	30.2	45.2	26.4	56.2	36.2	60.6
Monounsaturated FA	C18:1	23.8	15.2	20.0	10.8	25.3	10.4	25.6	17.4
Polyunsaturated FA	C18:2	2.5	2.9	5.3	4.89	2.2	7.0	7.1	9.6

Source: Adapted from Barman et al. 2012

(C18:0), oleic acids (C18:1) (Klass 1998), linoleic (C18:2), and linolineic acids (C18:3). The FA can be classified in medium chain (C10–C14), long chain (C16–C18) and very long chain species (> C20), and FA derivates. However, under unfavorable environment condition, many algae alter their lipid biosynthetic pathways to the formation and accumulation of neutral lipid (20–50% dry weight), mainly in the form of tryglycerides (TAGs). For biodiesel production, these neutral lipids have to be extracted from microalgae biomass (Alcaine 2010).

Lipid can be subdivided into two main groups: (1) the storage lipid (neutral or non-polar), and (2) structural (membrane or polar). The group of neutral lipid is formed by triacylglycerols or triglycerides (TAGs), steryl esters (SEs), and wax esters (WEs) (Lang 2007). TAGs generally serve as energy storage in microalgae that, once extracted, can be converted into biodiesel through transesterification reactions.

This neutral lipid bears a common structure of triple ester, where usually three long-chain fatty acids (FA) are coupled to a glycerol molecule. Transesterification displaces glycerol with small alcohols or methanol. Structural lipids typically have a high content of poly-unsaturated fatty acids (PUFAs), which are also essential nutrients for aquatic animals and humans. Polar lipid (phospholipid) and sterol are important structural components of cell membranes, which act as a selective permeable barrier for cell and organelles (Sharma et al. 2012).

Many algae are exceedingly rich in oil. The oil content of some microalgae exceeds 80% of the dry weight (DW) of algae biomass (Patil et al. 2008, Christi 2007). According to Shay (1993), algae can accumulate up to 65% of their total biomass as lipid, but according to Oilgae, some algae have only about 15–40% (DW), whereas palm kernel has about 50%, copra has about 60%, and sun flower has about 55%. Oil content itself can be estimated to be 64.4% of total lipid component (Hill and Feinberg 1984). In fact, microalgae have the highest oil yield among various plant oils. It can produce up to 100 tones oil/ha/yr, whereas palm, coconut, castor, and sun flower produce up to 5950, 2689, 1413, and 952 L/ha/yr, respectively (Oilgae).

Microalgal oil can be produced through either biological conversion to lipid, hydrocarbon or thermochemical liquefaction of algal cells. Direct extraction of microalgal lipid appears to be a more efficient methodology for obtaining energy from these organisms, than fermentation of algal biomass to produce either methane or ethanol (FAO). Alga cells of *Dunaliella tertiolecta* with moisture content of 78.4 wt% were converted directly into oil by the chemical liquefaction at around 300°C and 10 MPa. The oil yield was about 37% (organic basic). The oil obtained at a reaction temperature of 340°C and holding time 60 min, had viscosity of 150–330 mPas and heating value of 36 MJ/kg (Minowa et al. 1995).

3.1 Lipid Profile

Properties of biodiesel can be predicted from FA profile of lipid feed stock, therefore searching for promising algal genera is needed. Total lipid content and FA profile of freshwater, brackish water, and marine algae taxa, collected from Sundarban area was reported by Barman et al. (2012). Table 3.2 represents the total lipid and FA profile adapted from Barman et al. (2012), but just selected the major FA such as C16:0, C18:1, C18:2 of fresh water, brackish water, and marine water microalgae.

3.2 Optimization of Algal Biofuel Production

Algae cultivation has four basic requirements, i.e., carbon dioxide (CO_2), water, light, and space. By optimizing the quantity and quality of these requirements, it is possible to maximize the quantity of oil rich biomass. The flue gases from the power plants of industries are rich in CO_2 that would normally be released directly into the atmosphere, and thereby contribute to global warming. CO_2 can be used as a fertilizer in algae cultivation. Water with essential salts and minerals is needed for the growth of algae. Waste water from domestic or industrial sources usually already contains nitrogen and phosphate salts. The waste water can be added to the algal growth media directly. Light from the sun or from artificial sources is needed for the process of photosynthesis in open ponds or closed photobioreactors.

Full-spectrum light, about half of which is photosynthetically useful (400–700 nm), is normally used for microalgal growth; however, it has already been recognized that blue (420–450 nm) and red (660–700 nm) light are as efficient for photosynthesis as the full spectrum (Gouveia 2011). The cultivation of microalgae needs space; however, there are depth limitations. The ponds are kept shallow because of the need to keep the algae exposed to sunlight, and the limited depth to which sunlight can penetrate the pond's water. Race way ponds are about 15–35 cm deep to ensure adequate exposure to sunlight. Tubular photobioreactor has tubes of small diameter (0.2 m or less) to allow light penetration to the center of tube (Demirbas 2010).

Microalgae is typically grown under autotrophic condition, where algae utilizes sunlight as an energy source and carbon dioxide as the carbon source. Certain species of algae can also grow under heterotrophic condition. In heterotrophic growth, algae utilize a reduced organic compound as their carbon and energy source (Drapcho et al. 2008). Miao and Wu (2004) studied green algae *Auxenochlorella protothecoides* (formerly *Chlorella protothecoides*), and found that lipid content in heterotrophic condition could be as high as 55%, which was 4 times higher than in autotrophic conditions at 15%, under similar conditions. *Chlorella vulgaris* showed an increase in biomass production when the organism was grown under mixotrophic condition.

Mixotrophic is performing photosynthesis as the main energy source, though both organic compound and CO_2 are essential.

High lipid productivity of dominant, fast-growing algae is a major prerequisite for commercial production of microalgal oil derived biodiesel. However, under optimal growth conditions, large amounts of algal biomass are produced, but with relatively low lipid contents, while species with high lipid contents are typically slow growing. Major advances in this area can be made through the induction of lipid biosynthesis, e.g., by environmental stresses. There have been a wide range of studies carried out to identify and develop efficient lipid induction techniques in microalgae such as nutrient stress (e.g., nitrogen and/or phosphorus starvation), osmotic stress, radiation, pH, temperature, heavy metals, and other chemicals. In addition, several genetic strategies for increased triacylglycerides production and inducibility are currently being developed (Sharma et al. 2012).

Stress growth conditions can often be used to increase the formation of natural lipid. The stress was caused by either the use of nutrient-deficient media, or the addition of excess salt to nutrient-enriched media. The combination of both nutrient deficiency and salt enrichment appears to enhance lipid formation with *Isochrysis* sp., but to reduce it with *Dunaliella salina*. Interestingly, the free glycerol content can apparently be quite higher for *Dunaliella* sp. *Botryococcus braunii* exhibited relatively high lipid content under each set of growth conditions, but the highest was 54.2% (DW) under nutrient-deficient growth conditions (Klass 1998).

Nannochloropis sp. grown in media under nitrogen limited conditions will increase the lipid content from 28 to over 50%. Lipid productivity was demonstrated to have maximum of 150 mg/L day at 5 to 6% cell nitrogen, and was the initial nitrogen concentration exceeded 25 mg/l (Klass 1998). According to Shiffrin and Chrisholm in Weldy and Huesemann *Microchloropsis salina* (formerly *Monallantus salina*) produced as much as 72% lipid in nitrogen-deficient condition (Weldy and Huesemann 2013). Weldy and Huesemann argued that for lipid production, the percentage lipid content of microalgae was less important than the maximization of growth rates. The conclusion was Nitrogen-deficient culture will develop higher lipid content than N-sufficient cultures. Cultures under high light will develop higher lipid content than under low light, and culture lipid concentration and lipid production rates indicate that higher amounts of lipid are produced under N-sufficient condition and high light, due to higher biomass growth.

Xin et al. (2010) compared the freshwater microalgae *Scenedesmus* sp. with 11 species of high lipid content (reported by other researchers), and reported that *Scenedesmus* sp. showed the best ability to adapt to growth in secondary effluent, and had the highest microalgal biomass of 0.11 g/L (DW) and lipid content at 31–33%.

Benemann et al. (2013) reported their study on eight strains of microalgae. All microalgae were subjected to nitrogen limitation in batch cultures. Experiments were carried out primarily at one light intensity (300 microeinstein/m/sec), one CO_2 level (1% in air), and one N level (1.6 mM N-NO_3). *Chlorella* sp. had low lipid content under N sufficient and exhibited no significant increase in its lipid content, while overall productivity decreased rapidly upon N limitation. In contrast, *Nannochloropsis* sp. exhibited a relatively high lipid content under N sufficient conditions (25% of as free dry weight), a further increase upon N limitation (to over 50%), and a sustained high productivity after N limitation was induced. The other strains studied were intermediate in response. *Cyclotella* sp. exhibited a marked rise in lipid content upon N limitation (for 14 to 40%), and for a short period, relatively high lipid productivities in response to N limitation. *Chaetoceros gracilis* and *Isochrysis galbana* did not exceed 30% lipid contents—even after prolonged N deficiency. *Thalassiosira pseudonana, Ankistrodesmus falcatus,* and *Baekolavia* sp. exhibited maximum contents of 35 to 39%, however, lipid productivities were rather low.

4. Calculation of Biodiesel Production

Transesterification is a chemical process whereby ester is reacted with alcohol to form another ester. Figures 3.4 and 3.5 are transesterification reactions of biodiesel production process. The primary input is oil that has previously been extracted from microalgae, where R_1, R_2 and R_3 are long hydrocarbon chains, sometimes called fatty acid chains (Gerpen et al. 2007, Knothe et al. 2003, Gerpen et al. 2004, Gerpen and Knothe 2005).

There are three main types of FA that can be present in a triglyceride, i.e., saturated (Cn:0), mono-unsaturated (Cn:1), and poly-unsaturated with two or three double bonds (Cn:2,3), see Table 3.2. From the FA methyl ester composition, the cetane number (CN), kinematic viscosity, density, and heating value of biodiesel can be predicted (Verduzco et al. 2012). CN for biodiesel should be a minimum of 51. Low CN have been associated with more highly unsaturated components, such as ester of linoleic (C18:2) and linolenic (C18:3) acids (Knothe et al. 2003). According to Knothe (2008), ideal biodiesel feedstock would be composed entirely of C16:1 and C18:1.

As recommended by Schenk et al. 2008, a good quality of biodiesel should have fatty acid of C16:1 (palmitoleic), C18:1 (oleic), C14:0 (palmitic) with ratio of 5:4:1. The greatest content is palmitoleic acid with chemical structure: $CH_3(CH_2)_5 CH = CH(CH_2)_7 COOH$. For simplicity, consider the microalgae oil to consist of pure tripalmitolein. The tripalmitolein is a triglyceride in which all three fatty acid chains are palmitoleic acid. The tripalmitolein is reacted with methanol, the reaction will be that shown in Fig. 3.5. Note that the weights for each of the compounds in the reaction

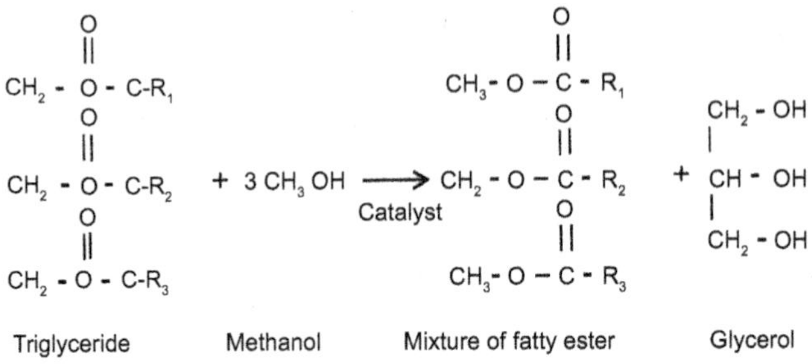

$$
\begin{array}{l}
\text{CH}_2 - \text{O} - \overset{\overset{\textstyle O}{\|}}{\text{C}}\text{-R}_1 \\[4pt]
\text{CH}_2 - \text{O} - \overset{\overset{\textstyle O}{\|}}{\text{C}}\text{-R}_2 \quad + \; 3\,\text{CH}_3\,\text{OH} \;\longrightarrow \\[4pt]
\text{CH}_2 - \text{O} - \overset{\overset{\textstyle O}{\|}}{\text{C}}\text{-R}_3 \\
\end{array}
\qquad
\begin{array}{l}
\text{CH}_3 - \text{O} - \overset{\overset{\textstyle O}{\|}}{\text{C}} - \text{R}_1 \\[4pt]
\text{CH}_2 - \text{O} - \overset{\overset{\textstyle O}{\|}}{\text{C}} - \text{R}_2 \quad + \\[4pt]
\text{CH}_3 - \text{O} - \overset{\overset{\textstyle O}{\|}}{\text{C}} - \text{R}_3 \\
\end{array}
\qquad
\begin{array}{l}
\text{CH}_2 - \text{OH} \\
\text{CH} - \text{OH} \\
\text{CH}_2 - \text{OH} \\
\end{array}
$$

| Triglyceride | Methanol | Mixture of fatty ester | Glycerol |

Figure 3.4. Transesterification reaction.

$$
\begin{array}{l}
\text{CH}_2\text{-O-C-CH}_3(\text{CH}_2)_3\text{CH=CH(CH}_3)_2\text{COOH} \\
\text{CH}_2\text{-O-C-CH}_3(\text{CH}_2)_3\text{CH=CH(CH}_3)_2\text{COOH} + 3\text{CH}_3\text{OH} \rightarrow \\
\text{CH}_2\text{-O-C-CH}_3(\text{CH}_2)_3\text{CH=CH(CH}_3)_2\text{COOH} \\
\end{array}
$$

Tripalmitolein 891.15 g + Methanol 96.12 g ➜ Methyl palmitioleate 895.28 g + Glycerol 92.10 g

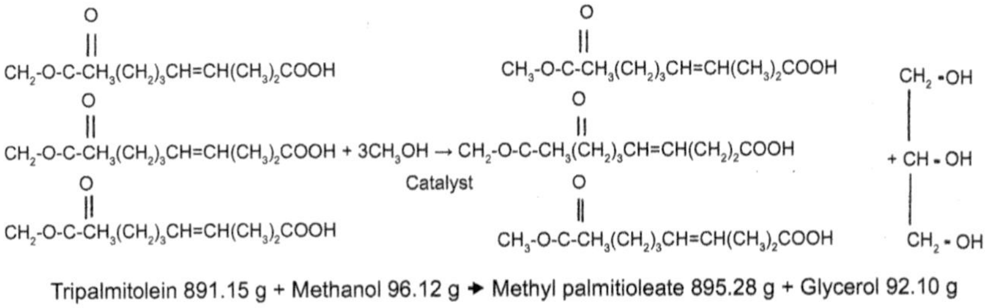

Figure 3.5. Transesterification of tripalmitolein.

are given. These are based on the fact that one molecule of tripalmitolein reacts with 3 molecules of methanol to produce 3 molecules of methyl palmitioleate or biodiesel and one mole of glycerol.

The molecular weight of tripalmitolein can be calculated by counting the number of carbons in the molecule (51), multiplying this by 12.0111, and also doing the same thing for hydrogen and oxygen:

C: 51 x 12.0111 = 612.566

H: 86 x 1.00797 = 86.6854

O: 12 x 16.0000 = 192.0000

Total mol.weight = 891.2515 g per mole

The molecular weight of methanol is 3 x 32.94 = 96.12 g and methyl palmitioleate is 3 x 298.4278 = 895.2834 g, and glycerol is 92.10 g.

Theoretically, transesterification reaction requires 3 moles of alcohol for each mole of oil. However, in practice, the molar ratio should be higher than that of stoichiometric ratio in order to drive the reaction towards completion, or the reaction is usually conducted by excess methanol. For example, below, the reaction requires 6 moles of alcohol (100% excess methanol) for each mole of oil (molar ratio 6:1), and the result of reaction of tripalmitolein with 100% excess methanol is shown in the following:

Tripalmitolein + 2 x Methanol ➔ Methyl + Glycerol + Excess
palmitioleate methanol

891.2515 g + 192.24 g ➔ 895.2834 g + 92.10 g + 96.12 g

On a volume basis, the reaction becomes:
100 liters of oil + 25.338 liters methanol ➔ 106.73 liters methyl palmitioleate
+ 7.612 liters glycerol + 12.65 liters excess methanol.

Table 3.3. Densities of Biodiesel Reactants.

	Density, g/liter
Tripalmitolein	929
Methanol	791.4
Methyl palmitioleate	875
Glycerol	1261.3

5. Pretreatment/Esterification

Free fatty acids (FFA) of the extraction oil should be checked before being used as raw materials of biodiesel by using titration process, because after harvesting of the microalgae, the chemicals in biomass may be subjected to degradation induced by the process itself, and also by internal enzyme in algal cells hydrolyze cellular lipids into the FFA (Liu et al. 2011, Mata et al. 2010).

The FFA will react with the alkali catalyst to form soap and water. Therefore, additional catalyst must be added to compensate for the catalyst loss to soap. When the FFA level is above 5%, the soap will inhibit separation of the methyl ester and glycerol, and cause emulsion formation during the water washing. Therefore, it is necessary to first convert FFA to methyl ester in order to make the FFA contents lower. The low FFA of pretreated oil is transesterified with an alkali catalyst to convert triglycerides to methyl ester (Huang et al. 2010, Sakthivel et al. 2011).

The microalgae oil firstly is mixed with alcohol and pH indicator, the doing titration with mixture of NaOH and distilled water. FFA content can be calculated after that. Acid value of the oil is 2 times of FFA, and defined as the number of milligrams of potassium hydroxide required to neutralize the FFA present in 1 g of oil sample.

If the FFA content < 1%, it can be ignored. If the FFA level > 1%, it is possible to be processed by adding alkali catalyst. When FFA levels > 5%, FFA should be reduced to below 1%, otherwise, the production yield will be low. Ramadhas et al. (2005) reported that the yield of esterification process decreases considerably if FFA value is greater than 2%.

Biodiesel feedstock are classified based on their FFA content as follows (Kinast 2005): very low FFA content or refined oils (FFA < 1.5%); low FFA content (FFA< 4%); high FFA content (FFA ≥ 20%). If the FFA > 2.5%, the oil was converted to glycerides in a pretreatment process with methanol using anhydrous H_2SO_4 (catalyst) or in esterification process. If the FFA < 2.5%, the oil can be processed directly by using transesterification process. According to Tyson (2002) if the FFA of feedstock < 4% can be processed by using transesterification process, and if the FFA between 4 to 20% uses a two-step process (acid esterification and base transesterification), while the FFA feedstock > 50%, it may be cost effective to hydrolyze the oil into 100% FFA, then proceed with acid esterification.

Technical options when FFA equal or exceeds 4% (Tyson 2002):

- Remove FFA content with NaOH and centrifuge (caustic stripping).
- Convert FFA into methyl esters with acid esterification then proceed with transesterification.
- Convert feedstock into 100% FFA and then proceed by acid esterification.
- Separate the FFA and triglycerides and treat separately.

To reduce FFA level, the following things can be done (Figs. 3.6 and 3.7). At first, the FFA content in the feedstock must be measured. Every one gram of FFA, 2.25 g of methanol and 0.05 g of H_2SO_4 should be added. Methanol is mixed with H_2SO_4, and then added to the feedstock, stirred for one hour with the temperature 60–65°C. Let the mixture settle so that a mixture of methanol and water will rise to the top, and the underlying layer is H_2SO_4 and oil. Decant the mixture of methanol, water, and H_2SO_4, then measure the new FFA content. If the FFA levels is still > 0.5%, the process is repeated again with a new FFA content. If the FFA content < 0.5%, then proceed to transesterification process (Gerpen et al. 2004).

An alternate procedure for processing high FFA is to hydrolyze the feedstock into pure FFA and glycerin. Typically, this is done in counter current reactor using sulfuric/sulfonic acid and steam. The output is pure FFA and glycerin (Gerpen et al. 2004).

An acid catalyst such as sulfuric acid can be used to esterify the FFA to methyl ester, as shown in the following reaction:

$$R\text{-}COOH \quad + \quad CH_3OH \quad \longleftrightarrow \quad R\text{=}COOCH_3 \quad + \quad H_2O \quad (1)$$

Fatty acid + Methanol   Acid Catalyst   Methyl Ester + Water

During the esterification process, water would be produced. When water content increases and accumulates, the esterification process will be disrupted or can even stop the reaction before completion (Gerpen and Knothe 2005). Lu et al. (2009) found that the conversion dropped from 95.6% to 52.2% when the water content increased from 0.0 to 1.5%. The dramatic

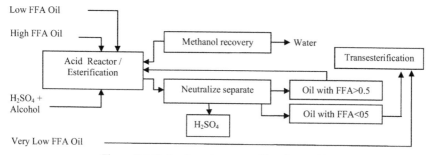

Figure 3.6. Pretreatment process of free fatty acid.

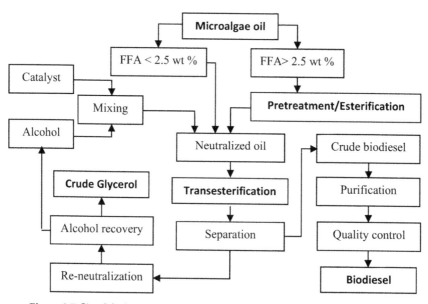

Figure 3.7. Simplified process flow chart biodiesel production (Leung et al. 2010).

drop of conversion with moisture content was caused by kinetic influence rather than the equilibrium influence. The water might greatly reduce the acid catalyst concentration in oil.

Many pretreatment methods for reducing the high FFA content of the oils have been proposed, including steam distillation, extraction by alcohol, and esterification by acid catalyst. Compared with the two former methods, esterification by acid catalysis makes the best use of the FFA in the oil and transforms it into biodiesel (Leung et al. 2010). Diaz-Felix et al. (2009) demonstrated that to reduce the FFA, 19.2% of yellow grease, with molar ratio of 9:1, 10% w/w sulfuric acid (based upon the FFA weight) were used. The entire conversion occurred during the first 30 min of the reaction.

The lack of conversion at later times may be due to accumulation of water released by the esterification of FFA that would reduce the reaction rate over time. This behavior is in accordance with that obtained by Ghadge and Raheman (2005), Canakci and Gerpen (2001).

Lu et al. (2009) used a two-step process consisting of esterification and transesterification to produce biodiesel from crude *Jatropha curcas* L. oil. Acid value of the oil was reduced from the initial 14 mg KOH/g oil to below 1.0 mg KOH/g oil in 2 h under condition of 12 wt% methanol, 1 wt% H_2SO_4 in oil at 70°C. The conversion of FFA was higher than 97% at 90°C in 2 h using 4 wt% solid acid and molar ratio of methanol to oil of 20:1. The esterification process was conducted in a 250 mL three neck flask. The flask was equipped with a mechanical agitator and a reflux condenser, and heated with a water bath to control the reaction temperature. In the experiments, flasks loaded with *Jatropha curcas* L. oil samples were firstly heated to the designated temperature, followed by the addition of the methanol and sulfuric acid mixture before turning on the agitator. The esterification product was separated in a tap funnel to obtain the upper oil layer, which was then washed with water several times until pH of washing water was close to 7.0. The oil was dried by anhydrous magnesium sulfate before subsequent transesterification.

Lu et al. (2009) also used the solid acid catalyst—the metatitanic acid. The metatitanic acid firstly was dried at 110°C in air for 5 h, and then crushed and sieved. The particles with a size under 125 μm were further calcined to prepare the solid acid catalyst. The esterification process took place inside an autoclave. Both the reactants and catalyst were added at the beginning, and the reactor was rapidly heated at 7°C/min under mechanical agitation of 1500 rpm. After completing the reaction, the reactor was quenched to stop the reaction. The slurry was filtered under vacuum and the liquid phase was allowed to settle in a tap funnel to separate the acidic water and oil phase. The acidic water and methanol were components in the upper layer. The oil phase was obtained at the lower layer, and was kept at 110°C for 90 min in an oven to evaporate the residual moisture and methanol. The treated oil was then used as the feedstock for transesterification process. The yield of biodiesel by transesterification was higher than 98% in 20 min using 1.3% KOH as catalyst and molar ratio methanol to oil 6:1 at 64°C.

Ramadhas et al. (2005) used high FFA rubber seed oil to biodiesel production. The FFA content of unrefined rubber seed oil was about 17%. They used a two-step process to convert the high FFA oil to biodiesel. The first step, acid catalyzed esterification, reduces the FFA content to less than 2%. The second step, alkaline catalyzed transesterification process, converts the product of the first step to biodiesel and glycerol.

In the first step, one liter of crude rubber seed oil requires 200 mL of methanol for the acid esterification process. The rubber seed oil is poured into the flask and heated to about 50°C. The methanol was added with the preheated oil and stirred for a few minutes. Sulfuric acid at 0.5% was also added with the mixture. Heating and stirring was continued for 20 to 30 min at atmospheric pressure. On completion of this reaction, the product was poured into a separating funnel for separating the excess alcohol. The excess alcohol, with sulfuric acid and impurities moves to the top surface, while the lower layer was refined oil for further processing.

The maximum conversion efficiency was achieved very close to the molar ration of 6:1. With further increase in molar ratio, there is only little improvement in the conversion efficiency. The amount of acid catalyst which was used in the process also affected the conversion efficiency of the process. The acid catalyst process attains the maximum conversion efficiency at 0.5% of sulfuric acid (Ramadhas et al. 2005).

The variables affecting the esterification process are methanol to microalgae oil molar ratio, acid catalyst concentration, temperature, and time (Venkanna and Reddy 2009). When the esterification process was complete, the product was transferred to a separating funnel, where the excess methanol along with impurities moved to the top layer and was removed. The bottom layer was used for the alkali transesterification.

Other pretreatment methods have been proposed for the high FFA content of the oil, such as steam distillation, and extraction by alcohol. Steam distillation for reducing high FFA required a high temperature and has low efficiency. However, extraction by alcohol method need a large amount of solvent, and the process is complicated. Compared with the two other treatment processes above, esterification is the best (Leung et al. 2010). If the FFA is very high, one step esterification pretreatment may not enough to reduce the FFA because of the high content of water produced during the reaction. The water must be removed by separation funnel before the next esterification process. Some researchers reduce the FFA by using acidic ion exchange resins in a packed bed, but the loss of the catalyst activity may be a problem. Other researchers use iodine as a catalyst. The iodine catalyst can be recycled after esterification reaction.

Ghadge and Raheman (2005) produced biodiesel from mahua oil (*Madhuca indica*) having high FFA. The initial FFA level of 19% (or 38 mg KOH/g) was reduced to less than 1% by a two-step pretreatment. The first optimum step was carried out with 0.35 v/v methanol to oil ratio in the presence of 1% v/v H_2SO_4 as an acid catalyst in 1 h reaction at 60°C and acid value reduced from 38 to 4.84 mg KOH/g. By using 0.20 v/v methanol to oil ratio (6:1 molar ratio), acid value could not go below 10 mg KOH/g even after 2 h reaction during the first step. In the second step, methanol to oil ratio of 0.30 v/v was found to be optimum to reduce the acid value below 2 mg KOH/g in 1 h reaction time. This was because some of the FFAs were already esterified during the first step and less amount of water was produced during the reaction.

The work of Canakci and Gerpen (2001) showed that the acid value of high FFA feedstocks could be reduced to less than 1% with a two-step pretreatment reaction. They used soybean oil added with 20 and 40% (by weight) palmitic acid. After addition, the acid value increased to 41.33 and 91.73 mg KOH/g for 20 and 40% palmitic acid, respectively. To investigate the influence of catalyst amount and reaction time on acid value, four different sulfuric acid amounts (0, 5, 15, and 25 wt% of the FFA), 1 h reaction time and molar ratio of 9:1 were selected. After 1 h, the mixture was allowed to settle. A methanol-water mixture separated from the oil phase at the top of the separatory funnel for the 5 wt% catalyst. However, for the 15 and 25 wt% catalyst, the methanol-water mixture was found at the bottom of the funnel. The acid value of the 20% palmitic acid mixture was reduced from 41.33 to 1.77 mg KOH/g when 5% sulfuric acid catalyst was used. When the catalyst level was increased to 25%, the acid value decreased to 0.54 mg KOH/g.

For the 40% palmitic acid case, the acid value decreased from 91.73 to as low as 6.25 mg KOH/g. However, even when 25% H_2SO_4 was used, the acid value did not reach to 2 mg KOH/g. In this case, when the FFAs are converted to ester, water formation inhibits the reaction. Therefore, water must be removed, and this could be accomplished using multi step process until the acid value reached below 1.0 mg KOH/g.

One major disadvantage of acid pretreatments (esterification) of high FFA oils is that they require large amounts of methanol in order for the reaction to reach a high yield. In addition, the performance of the esterification of FFA is remarkably different when using low FFA oils than high FFA oils (Diaz-Felix et al. 2009).

6. Transesterification

Transesterification process has been widely used to reduce the high viscosity of triglycerides. Transesterification is one of the reversible reactions and proceeds essentially by mixing the reactants. However, the presence of catalyst (acid or base) accelerates the conversion (Meher et al. 2006).

Raw materials with low FFA levels (less than 1%) can be directly processed with a single stage process method, called transesterification process. The feedstock is mixed with alcohol and alkali catalyst to produce biodiesel. As catalyst can be used sodium hydroxide catalyst 1% of triglyceride weight, or potassium hydroxide 1%, or sodium methoxide 0.25% (Gerpen et al. 2004).

Transesterification is a chemical reaction between triglyceride and alcohol in the presence of a catalyst. It consists of a sequence of three consecutive reversible reactions where triglycerides are converted to diglycerides, and then diglycerides are converted to monoglycerides,

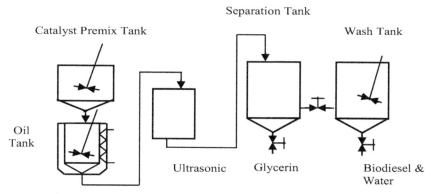

Figure 3.8. Schematic transesterification process of biodiesel production.

followed by the conversion of monoglycerides to glycerol. In each step an ester is produced and thus, three ester molecules are produced from one molecule of triglyceride (Sharma and Singh 2008).

Figure 3.8 shows the schematic transesterification process of biodiesel production (Amin 2009). The first step is removing water content from the oil by increasing its temperature up to 70°C for about 5–10 min. After that, it is allowed to cool, and by using a catalyst tank with mixed sodium hydroxide and methanol and stirring, sodium methoxide is produced. Meanwhile, clean oil is heated to 60°C for 5 min, mixed with the sodium methoxide, and the mixture transferred to ultrasonic or mixer equipment. This equipment agitates the solution for 30 min. After the mixing process, the solution is allowed to cool and separate. The separation process takes approximately from 15 to 60 min. The methyl ester (ME) or biodiesel would float on the top layer, while the denser glycerin would be in the bottom layer. In the last step, the biodiesel is washed, dried, and then quality tested. Figure 3.9 show process flow diagram of biodiesel plant.

The process of transesterification is affected by various factors depending upon the condition used. The effects of these factors are described below.

6.1 Effect of FFA and Moisture

The FFA should be lower than 3%. When the reactions do not meet this requirement, ester yield are significantly reduced. If the oils with high FFA, they must be refined by saponification using NaOH solution to remove the FFA. The addition of more sodium hydroxide catalyst compensates for higher acidity, but the resulting soap causes an increase in viscosity or formation of gels that interferes the reaction. The sodium hydroxide, potassium hydroxide, and sodium methoxide should be maintained in anhydrous state. Prolonged contact with air will diminish the effectiveness

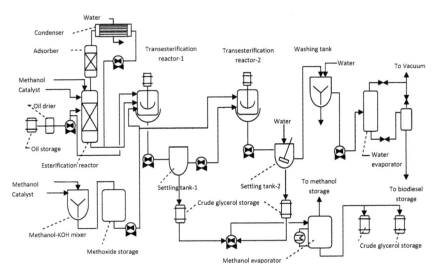

Figure 3.9. Process flow diagram of biodiesel plant.

of these catalysts through interaction with moisture and carbon dioxide (Meher et al. 2006). Moisture content in the catalyst will also disturb the esterification and transesterification reaction. As the initial water content increased, methyl ester content gradually decreased even when the initial water content was 1% of oil, water hindered the esterification, accelerating the inverse reaction (Park et al. 2010).

6.2 Catalyst Type and Concentration

Catalysts used for the transesterification of triglycerides are classified as alkali, acid, enzyme, or heterogeneous catalyst. Alkali catalysts like sodium hydroxide, sodium methoxide, potassium hydroxide, potassium methoxide are more effective, but mostly sodium hydroxides or potassium hydroxide have been used, both in concentration from 0.4 to 2 w/w% of the oil. Refined and crude oils with 1% either sodium or potassium hydroxide catalyst resulted successful conversion (Meher et al. 2006). When the FFA level less than 1%, common catalyst amounts are (Gerpen et al. 2004):

- Sodium hydroxide: 1% of triglycerides weight.
- Potassium hydroxide: 1% of triglycerides weight.
- Sodium methoxide: 0.25% of triglycerides weight.

When the FFA levels are above 1%, the amounts of catalyst can be calculated by following formulas (Gerpen et al. 2004):

- Sodium hydroxide: [% FFA] (0.144) + 1%.
- Potassium hydroxide: [% FFA] (0.197)/0.86 + 1%.
- Sodium methoxide: [% FFA] (0.190) + 0.25%.

For example, when adding sodium methoxide to feedstock with 1.5% FFA, the amount of catalyst would be: (1.5) (0.190) + 0.25% = 0.54% of the triglycerides weight. Note that a factor of 0.86 has been included with the potassium hydroxide calculation to reflect that reagent grade KOH is only 86% pure. If other grades of catalyst are used, this factor should be adjusted to their actual purity.

6.3 Molar Ratio of Alcohol to Oil

Transesterification is an equilibrium reaction, in which a large excess of alcohol is required to drive the reaction to the right. For maximum conversion to the ester, a molar ratio of 6:1 should be used (Meher et al. 2006). The amount of methanol/triglycerides molar ratio for base catalyzed transesterification stoichiometrically required is 3:1. However, transesterification reaction required excess alcohol. According to the research by Sahoo and Das (2009), by using Polanga oil, the minimum conversion can be achieved, if the volumetric ratio of oil to methanol is 12:1. The yield of ester is 85%. With further increase in volumetric ratio, there was no improvement in the conversion efficiency. Also, it has been found that the reduction in viscosity increase with increase in volume of methanol in the mixture.

6.4 Effect of Reaction Duration

The result of the experiment by Sahoo and Das (2009) with Polanga oil was that about 2 hr of the reaction was sufficient for the completion of the base catalyst transesterification. *Jatropha curcas* had an initial acid value of 28 mg KOH/g and Karanja oil has an acid value of 36 mg KOH/g. Patil and Deng (2009) produced biodiesel from *Jatropha curcas* and Karanja by using the two-step method: acid esterification and transesterification. The maximum ester conversions for Karanja and *Jatropha curcas* oil were found to be 80 and 90%, respectively, at the methanol to oil molar ratio of 6:1. The acid catalyst process attained maximum yield for Karanja oil and *Jatropha curcas* oil at 1 and 0.5 wt% catalyst concentration, respectively. Results obtained from their experiment with *Jatropha curcas* oil revealed that 2 h for acid esterification and 2 h for alkali transesterification were sufficient for the completion of the reaction. For the Karanja oil, experiment revealed that 45 min for acid esterification and 30 min for alkali transesterification were enough for the completion of the reaction.

7. Other Production Technique

7.1 One Step Acid Catalyzed Processing

Zheng et al. (2006) produced biodiesel from waste frying oil by using one step acid catalyzed processing. They use very high molar ratio and temperature at 70 and 80°C. At 70°C with oil:methanol:acid molar ratio of 1:245:3.8, and at 80°C with molar ratio in the range 1:74:1.9 to 1:245:3.8. The transesterification was essentially a pseudo first order reaction as result of the large excess of methanol which drops the reaction to completion (99 ± 1% at 4 h). In the presence of the large excess methanol, FFAs present in the waste oil were very rapidly converted to methyl esters in the first few minutes under the above conditions. Little or no monoglycerides were detected during the course of the reaction, and diglycerides present in the initial waste oil were rapidly converted to biodiesel. Nevertheless, this method is difficult in practice because excess methanol recovery needs energy and high cost.

7.2 Two Stage Process (Transesterification-Transesterification)

Çaylı and Küsefoğlu (2008) found that single stage process gives low production yield (86%). In the two stage based catalyzed process, they found in biodiesel production from used cooking oils, the yield increased to 96%. They used waste cooking oil (with acid number 6.5) 1,000 g, 4.2 g NaOH and 140 mL in the first stage, and then add with 1.8 g NaOH and 60 mL MeOH in the second stage.

The FFA level of waste cooking oils is 2 to 7% (Gerpen et al. 2004), and acid value of the refined sunflower oil is 0.15 (Marchetti et al. 2007) or 0.075% of FFA. This indicates that the FFA content up to 7% can be processed with a two-stage transesterification.

7.3 Two Stage Process (Esterification-Transesterification)

If the FFA content of raw material is medium or high, researchers used two stage processes, because with single stage method will produce saponification which creates a serious problem of product separation and will ultimately substantially decrease methyl ester yield.

There are at least four techniques for converting the FFA to biodiesel (Gerpen et al. 2004). The first is enzymatic method, but this method requires expensive enzymes, therefore no one is using this method on a commercial scale. The second method is glycerolizing. This technique involves adding glycerol to the feedstock and heating to high temperature (200°C), usually with a catalyst such as zinc chloride. The third is acid catalysis method uses strong acid, the reaction of FFA to alcohol esters is relative fast, but the transesterification of the triglycerides is very slow, taking several days to complete. The fourth method is acid catalysis followed by alkali catalysis. High FFA content should be reduced first with the acid catalyst process, which called esterification. If the FFA content reaches < 2%, then processed by transesterification.

The choice of acid and alkali catalysts depends on the FFA content in the feedstock. FFA should not exceed a certain amount for transesterification. Canakci and Gerpen (1999, 2001) reported that transesterification was not feasible if FFA content in the oil was about 3%.

It has been found that the alkaline catalyzed transesterification process is not suitable to produce esters from unrefined oils. In order to reduce the acid value, the oil should be refined. Refining of vegetable oils increases the overall production cost of the biodiesel. Acid esterification is a typical method of producing biodiesel from high FFA oil according to Canakci and Gerpen (1999), but it requires more methanol, and it is also time consuming.

Chung et al. (2008) reported that when the amount of FFA in the feedstock exceeds 0.5%, they react with the homogeneous alkali catalysts, form unwanted soap as by-products and deactivate the catalyst. Furthermore, the soap will make the downstreaming separation and purification of the biodiesel more difficult.

To solve this problem, researchers proposed a two-step process (Zullaikah et al. 2005, Canakci and Gerpen 2001, Dorado et al. 2002, Talens et al. 2007). The two-step method is most commonly used to feedstock with high FFA. Canakci and Gerpen (2001) developed a two-step pretreatment process to reduce the FFA levels of yellow oil (< 15% FFA) and brown grease (> 15% FFA) to less than 1%.

The first step was the pre-esterification of FFA with methanol, which was catalyzed by liquid acids. By this pretreatment, the FFA content of the oil is reduced to less than 1%. The second step, alkaline-catalyzed transesterification process, converted the products of the first step to biodiesel and glycerol. Homogeneous acid catalysts showed better adaptability to FFA than base catalysts, and could catalyze esterification and transesterification simultaneously (Wang et al. 2006). The commonly used catalyst during acid esterification of neat oil is sulfuric acid (H_2SO_4) (Ramadhas 2005, Ghadge and Raheman 2005, Sharma and Singh 2008). For waste cooking oil, the catalyst used was again sulfuric acid (Zhang et al. 2003, Tashtoush et al. 2004), but in this case, they reported that the conversion was low.

Marchetti and Errazu (2008) showed that sulfuric acid was an attractive alternative to produce biodiesel by direct esterification of spent oil with high amounts of FFA compared with conventional technology using KOH as catalyst. The esterification process that has been studied shows that the amount of FFA was reduced from 10.684% to around 0.54% w/w. Solid acid catalysts have lower catalytic activity but higher stability, thus, they could be used for feedstock with large amount of FFA without catalyst deactivation (Lotero et al. 2005).

7.4 Triple Stage Process

Sahoo and Das (2009) reported that they reduce the levels of FFA polanga oil (*Calophyllum inophyllum*) from 22 to < 1% through three stages. In the first stage, they mixed a liter of polanga oil with methanol 350 mL/L, toluene 5 mL/L, and ortho phosporic acid 5 mL/L. Reaction temperature was 66°C for 2 h and 2 h settling time.

In the second stage, they mixed the polanga oil 1.155 L (product of the first stage) with methanol 75 mL/L, toluene 5 mL/L, H_2SO_4 6.5 mL/L. Reaction temperature was 55°C for 4 h and settling time was 2 h. In the third stage, they mixed the polanga oil 1.05 L (product of second stage) with methanol 120 mL/L, potassium hydroxide 9 g/L, H_2SO_4 6.5 mL/L. Reaction temperature was 66°C for 4 h and settling time was 13 h.

Venkanna and Reddy (2009) also used three-stage method for biodiesel production by using high FFA honne oil. The first stage was pretreatment process (acid esterification) and used H_2SO_4 catalyst. The reaction was conducted at 60 ± 1°C for 120 min. When the first stage was complete, the bottom layer of the product (oil) was transferred to second stage. The oil product of acid esterification was heated to the desired temperature before starting the reaction or second stage.

In the second stage, the alkali methoxide solution was added to oil product from first stage, stirred, and then allowed to settle under gravity for 8 hr. In the third stage, the upper layer of alkali transesterification product was mixed with petroleum ether. The ester and un-reacted oil readily mixed into the petroleum ether with glycerol as a separate layer. The lower layer of glycerol was removed. The upper layer was a mixture of ester and un-reacted oil. The upper layer was heated to 65°C to remove methanol. The product was a honne oil methyl ester or biodiesel.

7.5 Supercritical Methanol

Biodiesel is produced by transesterification of triglyceride use of alkali catalyst. In this method, FFAs react with alkali catalyst producing saponified products. The use of acid catalyst results in long reaction time and the process is still sensitive to water. Kyoto University used of supercritical methanol technology for biodiesel production. There are two type of supercritical method, i.e., the one step supercritical methanol method (Saka process) shown in Fig. 3.10, and the two-step method (Saka-Dadan process) shown in Fig. 3.11.

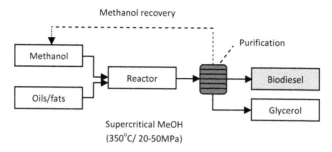

Figure 3.10. Schematic diagram of the one step supercritical methanol method.

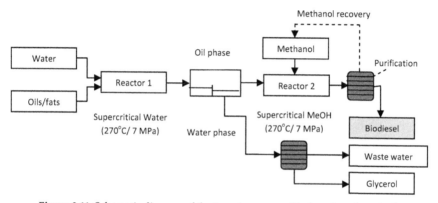

Figure 3.11. Schematic diagram of the two step supercritical methanol method.

Saka and Minami (2006) demonstrated that supercritical methanol has the ability to convert oils/fats into biodiesel without any catalyst. In the one step method, the optimum temperature was 350°C and 20 MPa, molar ratio methanol:oil = 42:1. In the two step method, subcritical water (270°C and 7 MPa) was used in the first reactor, and in the second reactor supercritical methanol (270°C and 7 MPa) was used. The one step method can produce a higher yield of biodiesel than the alkali catalyst method. Compared to the alkali catalyzed, the superiority of the one step method can be summarized: (i) the production process becomes much simpler, (ii) the reaction is very fast, (iii) FFA in oils/fats can be converted to biodiesel through methyl esterification, and (iv) the yield of biodiesel is higher. In the two-step method, oils/fats are first treated in subcritical water for hydrolysis reaction to produce fatty acid. After hydrolysis process, the reaction mixture was separated into oil phase and water phase by decantation. The oil phase (upper portion) was mainly fatty acid, while lower portion contains glycerol in water. The separated oil phase is then mixed with methanol and treated at supercritical condition to produce biodiesel.

7.6 In situ Transesterification

Wahlen et al. (2011) produce biodiesel from many microalgae. They developed approach for the direct production of biodiesel by *in situ* transesterification from high triglyceride accumulating microalgae, such as *Chaetoceros gracilis, Chlorella sorokiniana, Tetraselmis suecica,* etc. To determine the optimal condition for biodiesel production, experiments were conducted with 100 mg of lyophilized algal biomass. Methanol containing sulfuric acid as a catalyst was added to the reaction vessel containing a stir bar. Volume of methanol and the amount of sulfuric acid was varied to determine the amount necessary for optimal biodiesel production. Reactions were conducted in a commercial scientific microwave, where conditions of time and temperature could be controlled with precision. Once completed, reactions were stopped by addition of chloroform to the reaction vessel forming a single phase solution with the methanol. Phase separation was then accomplished by washing the methanol-chloroform solution with water, followed by centrifugation. The methanol and sulfuric acid partitioned with the water in the upper phase, while the methyl ester, triglycerides, and other lipids partitioned with chloroform in the lower, organic phase. The residual biomass formed a layer at the boundary between these two phases. The chloroform phase was removed with a gas tight syringe to a flask. The remaining biomass was washed twice with chloroform to recover residual methyl esters and lipids.

7.7 Non Catalytic Transesterification

Researchers Joelianingsih et al. (2008) developed a new reactor to produce fatty acid methyl ester by blowing bubbles of superheated methanol vapor continuously into vegetable oil without using any catalysts. Figure 3.12 is schematic flow diagram of reactor used in the experiment (Joelianingsih et al. 2008). The reactor was initially charged with the refined palm oil and heated to the desired temperature. Liquid MeOH was pump out of the dehydration column to the tin bath for vaporization. The MeOH vapor was taken through a ribbon heater and the reaction started by blowing the bubbles of superheated MeOH (0.1 MPa, 505–533 K) continuously into the reactor at fixed flow rate of 4 g/min. Reacted product in gas phase were condensed and collected using a glass container. The reaction products were taken from the glass container every 20 min and then weighed. The optimum reaction temperature which gives the highest methyl ester content (95.17% w/w) was 523 K, while the rate constant of the total system increased as the reaction temperature was increased.

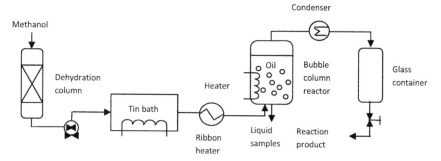

Figure 3.12. Flow diagram of reactor used in non catalytic transesterification experiment.

7.8 BIOX Co-Solvent Process

Due to poor methanol and oil miscibility, conversion of oil to biodiesel is very slow reaction. Use of a co-solvent that is soluble in both methanol and oil may improve reaction rate. The BIOX process is designed to overcome slow reaction time cause by the extremely low solubility of alcohol in the triglycerides phase. This process uses either tetrahydrofuran (THF) or methyl tert-butyl ether (MTBE) $C_5H_{12}O$ as a co-solvent to generate a one-phase system. The result is a fast reaction, on the order 5–10 min, and no catalyst residues in either the ester or the glycerol phase (Balat and Balat 2010). The excess methanol and co-solvent are recovered in single step after the reaction is complete.

BIOX's patented production process converts first the FFA (by way of acid esterification) and then the triglycerides (by way of transesterification), through the addition of a co-solvent, in a two-step, single phase, continuous process at atmospheric pressures and near-ambient temperatures (Demirbas 2009), all in less than 90 min, rather than the several hours or even days under conventional processes, with the highest possible conversion yields. The co-solvent is then recycled and reused continuously in the process (Anonymous bioxcorp 2015).

7.9 Membrane Based Separation

Miscibility is an important factor in biodiesel production. The conventional transesterification method results in a two phase reaction which is as a result of mass-transfer limited. More specifically, the vegetable oil and methanol are not miscible. The addition of co-solvent to generate a homogeneous reaction mixture can greatly enhance the reaction rate, but the co-solvent must eventually be separated from the biodiesel and this requires additional processing. Dube et al. (2007) used of membrane separation technology (Fig. 3.13), which exploit the immiscibility of the oil and methanol. Methanol

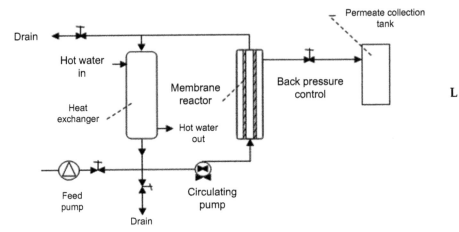

Figure 3.13. Schematic diagram of membrane reactor.

and sulfuric acid were pre-mixed and charged into membrane reaction system prior to each reaction. Canola oil was charged into the reactor and circulation pump was started. After 10 min circulation time, methanol and acid catalyst were charged continuously into reactor. The heat exchanger was switched on to achieve the reaction temperature. The reaction is using temperature of 65°C, 2 wt% catalyst, and 6.1 mL/min of flow rate (methanol/acid catalyst). The conversion is 64% (using acid-catalyst), and 96% (using base-catalyst).

8. Optimization of Biodiesel Production

A common method for establishing the biodiesel potential of an algal strain is to determine the total lipid content (Wahlen et al. 2011). A good quality of biodiesel should have fatty acids of C16:1, C18:1, C14:0 with ratio of 5:4:1 (Schenk et al. 2008). The problem is not many microalgae have like as recommended by Schenk. Islam et al. (2013) investigated nine microalgae species and found only one *Nannochloropsis oculata*, that was close to recommended ratio 5.1:3.5:1. From eight marine water microalgae species investigated by Barman et al. (2012), *Leptolyngbya tenuis* (formerly *Phormidium tenue*) was close to the recommended ratio of 5:3.16:1. To overcome the problem, it can be mixed some microalgae in order to reach the composition of fatty acid closest to Schenk recommendation.

Other problem in biodiesel product is the FFA oil content. If the FFA oil is more than 3%, the esterification reaction by using acid catalyst can be used to reduce the FFA, but the problem with acid catalyst is that the water production from the following reaction:

FFA + methanol → methyl ester + water

FFA can be removed by chemical neutralization or physical deacidification. Chemical neutralization involves treatment with NaOH, KOH, or by using strong acid or the esterification by using acid catalyst (Dunford 2014).

One approach is to simply add so much excess methanol during pretreatment that the water is diluted to the level where it does not limit the reaction. Molar ratio of alcohol to FFA as high as 40:1 may be needed. The disadvantage of this approach is that more energy is required to recover the excess methanol. Another approach would be to let the acid catalyzed esterification proceed as far as it will go until is stopped by water formation, the boil off the alcohol and water. If the FFA is still too high, the process can be continued for multiple stages and will use less volume methanol than the previous approach (Gerpen et al. 2004).

If the FFA oil content is less than 3%, it can be processed by using transesterification reaction or other production technique. Zheng et al. (2006) produce biodiesel from waste frying oil by using one step acid catalyzed processing and used high molar ratio of the oil to methanol 254:1. The recovery of excess methanol leads to high cost.

Supercritical methanol technology for biodiesel production used molar ratio methanol:oil = 42:1, temperature was 270 to 350°C and pressure 7 to 20 MPa. The inventors (Saka and Minami 2006) claimed that the superiority of the one step method (i) the production process become much simpler, (ii) the reaction is very fast, (iii) FFA in oils/fats can be converted to biodiesel through methyl esterification, and (iv) the yield of biodiesel is higher. The question is how high the cost production would be when this process is used in industry scale.

Wahlen et al. (2011) developed *in situ* transesterification from high triglyceride accumulating microalgae. They used methanol as solvent for extracting the lipids from biomass, and as a reactant for converting the lipids to biodiesel. The molar ratio of methanol:oil is 30:1. Fatty acid methyl ester yield were observed when the temperature was increase from 60 to 80°C for 20 min. Lin and Hsiao (2013) worked to optimize of biodiesel production from waste vegetable oil assisted by co-solvent and microwave and using two stages processes.

The following points are to be considered in the optimization process of biodiesel:

- Initial acid value to determine the stage of process.
- If using two stage processes, the first stage is usually H_2SO_4 as catalyst or using a solid acid. The amount of H_2SO_4 catalyst must be considered, because if it is too much, the color of biodiesel will be darker.
- The problem of water formation during esterification which can disturb the process.

- Molar ratio of alcohol/oil, alcohol recovery and cost production, production time, production yield, quality, and the problem of purification of the crude biodiesel should be also considered.

9. Summary

Microalgae have been suggested to be good candidates for fuel production because of their higher photosynthetic, efficiency, higher biomass production, and faster growth compared to other energy crops. Algae contain protein, carbohydrates and lipids. Lipids can be processed to biodiesel, carbohydrates to be ethanol and H_2, and proteins as raw material of biofertilizer. The difficulties in efficiency biodiesel production from microalgae lie not in the extraction of the oil, but in finding an algal strain with high lipid content and fast growth rate (Manivasagan and Se-Kwon Kim 2015). There are many ways to convert the oil and fats into biodiesel. There are transesterification, esterification, blending, micro emulsion and pyrolysis, but transesterification and esterification being the most commonly used method.

Other important factor in biodiesel production is fatty acids (FA) type, and its amount. There are three main type of the FA that can be present in a triglyceride, i.e., saturated, mono-unsaturated and poly-unsaturated with two or three double bonds. From the FA methyl ester composition, the cetane number (CN), kinematic viscosity, density, and higher heating value of biodiesel can be predicted.

Several important scientific and technical barriers remain to be overcome (Sánchez et al. 2011) before commercial reality. Although researchers have found many production processes, it cannot be applied commercially yet, because the production cost and energy consumption is still high.

Keywords: Biodiesel; microalgae; transesterification; esterification; oil extraction

References

Agarwal, A.K. 2007. Biofuel (alcohol and biodiesel) application as fuel for internal combustion engine. Prog. Energ. Comb. Scie. 33: 233–71.
Alcaine, A.A. 2010. Biodiesel from microalgae. Final degree project. Royal School of Technology Kungliga Tekniska Högskolan. Sweden.
Amin, S. 2007. Cara memproduksi biodiesel dari berbagai bahan baku nabati "Biodiesel production from vegetables feedstock". BPPT Press. Jakarta. Indonesia.
Amin, S. 2009. Review on biofuel oil and gas production process from microalgae. Energ. Conv. Manag. 50: 1834–40.
Austin, A. 2013. Open ponds versus close bioreactors. http://biomassmagazine. com/articles /3618/open-ponds-versus-close-bioreactors/. accessed 29 September 2013.
Balat, M. and H. Balat. 2010. Progress in biodiesel processing. Applied Energ. 87: 1815–35.

Barman, N., G.G. Satpati, S. Sen Roy, N. Khatoon, R. Sen, S. Kanjilat et al. 2012. Mapping algae of Sundarban origin as lipid feedstock for potential biodiesel application. Algal Biomass Utiliz. 3(2): 42–49.

Beal, C.M., R.E. Hebner, D. Romanovicz, C.C. Mayer and R. Connelly. 2013. Progression of lipid profile and cell structure in a research-scale production pathway for algal biocrude. Renew. Energ. 50: 86–93.

Benemann, J.R., D.M. Tillett, Y. Suen, J. Hubbard and T.G. Tornabene. 2013. Chemical profile of microalgae with emphasis on lipids. Final Report to the Solar Energy Research Institute. School of Applied Biology. George Institute of Technology. Atlanta. www.nrel.gov/docs/legosti/old115418.pdf/. accessed 5 November 2013.

Bioenergynet. 2013. Energy production from algae. Part 2: Cultivation system. http://www.bioenergynet.com/article/technology/innovation/424-energy-production-from -algae-part-2-cultivation-system/, accessed 29 September 2013.

Biox. 2015. How biodiesel is made. http://www.bioxcorp.com/what-is-biodiesel/how-biodiesel-is-made/. accessed 5 September 2015.

Borowitzka, M.A. 2013. Energy from microalgae: a short history. *In*: Borowitzka, M.A. and N.R. Moheimani (eds.). Algae for Biofuel and Energy. Development in Applied Phycology 5. Springer Science.

Brennan, L. and P. Owende. 2010. Biofuel from microalgae-A review of technologies for production processing, and extraction of biofuels and co-products. Renew. & Sustain. Energ. Review 14: 557–77.

Bridgwater, A.V., D. Meier and D. Radlein. 1999. An overview of fast pyrolysis of biomass. Org. Geochem. 30: 1479–93.

Bridgwater, A.V. and G.V.C. Peacocke. 2000. Fast pyrolysis processes for biomass. Renew. Sustain. Energ. Review 4: 1–73.

Campbell, M.N. 2008. Biodiesel: Algae as a renewable source for liquid fuel. Guelph. Eng. 1: 2–7.

Canakci, M. and J.V. Gerpen. 1999. Biodiesel production via acid catalysis. Trans ASAE 42(5): 1203–10.

Canakci, M. and J.V. Gerpen. 2001. Biodiesel production from oil and fats with high free fatty acid. Trans. ASAE 44(6): 1429–36.

Çaylı, G. and S. Küsefoğlu. 2008. Increased yields in biodiesel production from used cooking oils by a two-step process: Comparison with one step process by using TGA. Fuel Proc. Tech. 89(2): 118–22.

Christi, Y. 2007. Biodiesel from microalgae. Biotech. Adv. 25: 294–306.

Chung, K.H., D.R. Chang and B.G. Park. 2008. Removal of free fatty acid in waste frying oil by esterification with methanol on zeolite catalysts. Biores. Tech. 99: 7438–43.

Diaz-Felix, W., M.R. Riley, W. Zimmt and M. Kazz. 2009. Pretreatment of yellow grease for efficient production of fatty acid methyl esters. Biomass & Bioenerg. 33: 558–63.

Demirbas, A. 2009. Progress and recent trend in biodiesel fuels. Energ. Conv. Manag. 50: 14–34.

Demirbas, A. 2010. Use of algae as biofuel source. Energ. Conv. Manag. 51: 2738–49.

Demirbas, A. and M.F. Demirbas. 2010. Algae energy. Algae as a new source of biodiesel. pp. 97–133. Springer. London.

Dorado, M.P., E. Ballesteros, J.A. Almeida, C. Schellert, H.P. Lohrlein and R. Krause. 2002. An alkali-catalyzed transesterification process for high free fatty acid waste oils. Trans ASAE 45(3): 525–29.

Drapcho, C.M., N.P. Nhuan and T.H. Walker. 2008. Biofuels engineering process technology. pp. 197–259. McGraw Hill. New York.

Dube, M.A., A.Y. Tremblay and J. Liu. 2007. Biodiesel production using a membrane reactor. Biores. Tech. 98: 639–47.

Dunford, N.F. 2014. Biodiesel production techniques. Food Tech. Fact. Sheet 405-744-6071. FAPC 150: 1–4. www.fapc.biz/files/factsheets/faps 150.pdf/. accessed 16 January 2014.

FAO. 2007. Oil production. FAO Corp Doc Repository. www.fao.org/docrep/w7241e/w7241e0h.htm/. accessed 02 May 2007.

Ghadge, S.V. and H. Raheman. 2005. Biodiesel production from mahua (*Madhuca indica*) oil having high free fatty acids. Biomass & Bioenerg. 28: 601–05.

Gerpen, J.V., B. Shanks, R. Pruszko, D. Clements and G. Knothe. 2004. Biodiesel production technology. NREL/SR-510-36244.

Gerpen, J.V. and G. Knothe. 2005. Basic of the transesterification reaction. pp. 29–44. *In*: Knothe G., J.V. Gerpen and J. Krahl (eds.). The Biodiesel Handbook. AOCS Press. Champaign. Illinois.

Gerpen, J.V., C.L. Peterson and C.E. Goering. 2007. Biodiesel: an alternative fuel for compression ignition engines. ASAE Distinguished Lecture No. 31. Agric. Eq. Tech. Conference 11–14 February.

Gouveia, L. 2011. Microalgae as a feedstock for biofuels. Springer. Heidelberg.

Hill, A.M. and D.A. Feinberg. 1984. Fuel from microalgae lipid products. *In*: The Energy from Biomass: Building on a Generic Technology Base. Secon. Tech. Review Meeting. Portland-Oregon. 23–25 April.

Huang, G.H., F. Chen, D. Wei, X.W. Zhang and G. Chen. 2010. Biodiesel production by microalgal biotechnology. Applied Energ. 87: 38–46.

Islam, M.A., M. Magnusson, R.J. Brown, G.A. Ayoko, Md. N. Nabi and K. Heimann. 2013. Microalgae species selection for biodiesel production based on fuel properties derived from fatty acid profiles. Energies 6: 5676–5702.

Itoh, S., A. Suzuki, T. Nakamura and S. Yokoyama. 1994. Production of heavy oil from sewage sludge by direct thermochemical liquefaction. Desalination 98: 127–33.

Jain, A. and V.L. Sirisha. 2015. Algal biodiesel: Third-generation biofuel. pp. 423–457. *In*: Se-Kwon Kim and Choul-Gyun Lee (eds.). Marine Bioenergy. CRC. Press. Boca Raton. FL.

Janssen, M.G.J. 2002. Cultivation of microalgae: Effect of light/dark cycles on biomass yield. Ph.D. Thesis. Wageningen University. Ponsen & Looijen BV. Netherlands.

Joelianingsih, H. Maeda, S. Hagiwara, H. Nabetani, Y. Sagara, T.H. Soerawidjaya, A.H. Tambunan and K. Abdullah. 2008. Biodiesel fuels from palm oil via non catalytic transesterification in a bubble column reactor at atmospheric pressure: A kinetic study. Renew Energ. 33: 1629–36.

Kinast, J.A. 2005. Production of biodiesels from multiple feedstocks and properties of biodiesels and biodiesel/diesel blends. Final Report. NREL/SR-510-31460.

Klass, D.L. 1998. Biomass for renewable energy, fuels, and chemicals. Elesevier Academic Press. pp. 333–344.

Knothe, G., A.C. Matheus and T.W. Ryan III. 2003. Cetane number of branched and straight chain fatty esters determined in an ignition quality tester. Fuel 82: 971–75.

Knothe, G. 2005. The history of vegetable oil-based diesel fuel. pp. 8. *In*: Knothe, G., J.V. Gerpen and J. Krahl (eds.). The Biodiesel Handbook. AOCS Press. Champaign. Illinois.

Knothe, G. 2008. Designer biodiesel: Optimizing fatty ester composition to improve fuel properties. Energ. & Fuel 22: 1358–64.

Lang, I. 2007. New fatty acids, oxylipins and volatiles in microalgae. Ph.D. Dissertation. Georg-August-University. Göttingen.

Lee, J.V., C. Yoo, S.Y. Jun, C.Y. Ahn and H.M. Oh. 2010. Comparison of several methods for effective lipid extraction from microalgae. Biores. Tech. 101: 575–77.

Leung, D.Y.C., X. Wu and M.K.H. Leung. 2010. A review on biodiesel production using catalyzed transesterification. Applied Energ. 87: 1083–95.

Lin, C.C. and M.C. Hsiao. 2013. Optimization of biodiesel production from waste vegetable oil assisted by co-solvent and microwave using a two step process. Sustain Bioenerg. Sys. 3: 1–6.

Liu, J., J. Huang and F. Chen. 2011. Microalgae feedstock for biodiesel production. pp. 133–160. *In*: Stoychwa, M. (ed.). Biofuel Feedstock Processing Technologies. InTech. Rijeka Croatia.

Lotero, E., Y. Liu, D.E. Lopez, K. Suwannakarn, D.A. Bruce and J.G. Goodwin. 2005. Synthesis of biodiesel via acid catalysis. Ind. Eng. Chem. Res. 44: 5353–63.

Lu, H., Y. Liu, H. Zhou, Y. Yang, M. Chen and B. Liang. 2009. Production of biodiesel from *Jatropha curcas* L. oil. Comput. Chem. Eng. 33: 1091–96.

Manivasagan, P. and Se-Kwon Kim. 2015. Introduction to marine bioenergy. pp. 3–11. *In*: Se-Kwon Kim and Choul-Gyun Lee (eds.). Marine Bioenergy. CRC. Press. Boca Raton. FL.

Marchetti, J.M., V.V. Miguel and A.F. Errazu. 2007. Possible methods for biodiesel production. Renew Sustain Energ. Review 11: 1300–11.

Marchetti, J.M. and A.F. Errazu. 2008. Esterification of free fatty acids using sulfuric acid as catalyst in the presence of triglycerides. Biomass & Bioenerg. 32: 892–5.

Mata, T.M., A.A. Martins and N.S. Caetano. 2010. Microalgae for biodiesel production and other application. A review. Renew Sustain Energ. Review 14: 217–32.

McKendry, P. 2003. Energy production from biomass (part 2): Conversion technologies. Biores. Tech. 83: 47–54.

Miao, X. and Q. Wu. 2004. High yield bio-oil production from fast pyrolysis by metabolic controlling of *Chlorilla protothecoides*. Biotech. 110(1): 85–93.

Miao, X., Q. Wu and C. Yang. 2004. Fast pyrolysis of microalgae to produce renewable fuels. Anal. Applied Pyrolysis 71: 855–63.

Minowa, T., S. Yokoyama, M. Kishimoto and T. Okakurat. 1995. Oil production from algal cells of *Dunalliela tertiolecta* by direct thermochemical liquefaction. Fuel 74(12): 1735–38.

Meher, L.C., D.V. Sagar and S.N. Naik. 2006. Technical aspect of biodiesel production by transesterification—A review. Renew Sustain Energ. Review 10(3): 248–68.

Murakami, M., S. Yokoyama, T. Ogi and K. Koguchi. 1990. Direct liquefaction of activated sludge from aerobic treatment of effluents from the corn starch industry. Biomass 23: 215–28.

Oilgae. 2008. Algal oil yield. http://oilgae.com/algae/oil/yield/yield.html/. accessed 18 July 2008.

Ozkurt, I. 2009. Qualifying of safflower and algae for energy. Energy Education Science and Technology Part A Energy Science and Research 23: 145–151.

Park, J.Y., Z.M. Wang, O.K. Kim and J.S. Lee. 2010. Effects of water on the esterification of free fatty acids by acid catalysts. Renew Energ. 35: 614–18.

Patil, V., K.Q. Tran and H.R. Giselrod. 2008. Toward sustainable production of biofuels from microalgae. Int. J. Mol. Sci. 9: 1188–95.

Patil, P.D. and S. Deng. 2009. Optimization of biodiesel production from edible and non edible vegetable oils. Fuel 88: 1302–06.

Prabandono, K. and S. Amin. 2015a. Biofuel production from microalgae. pp. 145–158. *In*: Se-Kwon Kim (ed.). Handbook of Marine Microalgae. Elsevier. London.

Prabandono, K. and S. Amin. 2015b. Production of biomethane from microalgae. pp. 303–325. *In*: Se-Kwon Kim and Choul-Gyun Lee (eds.). Marine Bioenergy. CRC Press. Boca Raton. F.L.

Qi, Z., C. Lie, W. Tiejun and X. Ying. 2007. Review of biomass pyrolysis oil properties and upgrading research. Energ. Conv. Manag. 48: 87–92.

Raja, A., C. Vipin and A. Aiyappan. 2013. Biological importance of marine algae—An overview. Int. J. Curr. Microbiol. Applied. Sci. 2(5): 222–27.

Ramadhas, A.S., S. Jayaraj and C. Muraleedharan. 2005. Biodiesel production from high FFA rubber seed oil. Fuel 84: 335–40.

Sahoo, P.K. and L.M. Das. 2009. Process optimization for biodiesel from *Jatropha*, karanja and polanga oil. Fuel 84: 335–40.

Saka, S. and E. Minami. 2006. A novel non catalytic biodiesel production process by supercritical methanol as NEDO "High efficiency bioenergy conversion project". The 2nd Joint Int. Conf. on SEE. Bangkok Thailand.

Sakthivel, R., S. Elumalai and M.M. Arif. 2011. Microalgae lipid research, past, present: A critical review for biodiesel production, in the future. Exp. Sci 2(10): 29–48.

Sánchez, E., K. Ojeda, M. El-Halwaji and V. Kaparov. 2011. Biodiesel from microalgae oil production in two sequential esterification/transesterification reactors: pinch analysis of heat integration. Chem. Eng. J. 176-177: 211–16.

Sawayama, S., T. Minowa and S.Y. Yokoyama. 1999. Possibility of renewable energy production and CO_2 mitigation by thermochemical liquefaction of microalgae. Biomass Bioenerg. 17: 33–9.

Schenk, P.M., S.R. Thomas-Hall, E. Stephens, U.C. Marx, J.H. Mussgnug, C. Posten, O. Kruse and B. Hankamer. 2008. Second generation biofuels: High-efficiency microalgae for biodiesel Production. Bioenerg. Research 1: 20–43.

Shah, G.C., V. Pawar, M. Yadav and A. Tiwari. 2013. Comparision of different algal species for the higher production of biodiesel. Agr. Sci. Research 3(1): 23–28.

Sharma, K.K., H. Schuhmann and P.M. Schenk. 2012. High lipid induction in microalgae for biodiesel production. Energies 5: 1532–53.

Sharma, Y.C. and B. Singh. 2008. Development of biodiesel from karanja, a tree found in rural India. Fuel 87: 1740–2.

Shay, E.G. 1993. Diesel fuel from vegetable oils: Status and opportunities. Biomass & Bioenerg. 4(4): 227–242.

Sheechan, J., T. Dunahay, J. Banemann and P. Roessler. 1998. A look back at the US Depart of Energy's Aquatic Species Program-Biodiesel from algae. NREL/TP-580-24190. U.S. Depart of Energy's office of fuel development. Colorado.

Talens, L., G. Villalba and X. Gabarrell. 2007. Exergy analysis applied to biodiesel production. Resour. Conserv. Recycl. 51: 397–07.

Tashtoush, G.M., M.I. Al-Widyan and M.M. Al-Jarrah. 2004. Experimental study on evaluation and optimization of conversion of waste animal fat into biodiesel. Energ. Conver. Manga 45: 2697–711.

Tyson, K.S. 2002. Brown grease feedstocks for biodiesel, NREL 19 June. www. easternet. edu/depts/sustainenergy/calender/Tyson/Technology/Feedstocks.pdf/. accessed 18 November 2005.

Verduzco, L.F.R., J.E.R. Rodriguez and A.R.J. Jacob. 2012. Predicting cetane number, kinematic viscosity, density and higher heating value of biodiesel from its fatty acids methyl ester composition. Fuel 91: 102–11.

Venkanna, B.K. and C.V. Reddy. 2009. Biodiesel production and optimization from *Calophyllum inophyllum* linn oil (honne oil)—A three stage method. Biores. Tech. 100: 5122–25.

Wahlen, B.D., R.M. Willis and L.C. Seefeldt. 2011. Biodiesel production by simultaneous extraction and conversion of total lipids from microalgae, cyanobacteria, and wild mixed-cultures. Biores. Tech. 102: 2724–30.

Wang, Y., S. Ou, P. Liu, F. Xue and S. Tang. 2006. Comparison of two different processes to synthesize biodiesel by waste cooking oil. Mol. Catal A: Chem. 252: 107–12.

Weldy, C.S. and M. Huesemann. 2013. Lipid production by *Dunaliella salina* in batch culture: Effect of nitrogen limitation & light. US-DOE. Undergrad Research. www.scied.sceince. doe.gov/. accessed 1 November 2013.

Wikipedia. 2013. Pond and bioreactor cultivation methods. http://en.wikipedia.org /wiki/ Algaculture/. accessed 29 September 2013.

Wikipedia. 2013. Biodiesel. http://en.wikipedia.org/wiki/Biodiesel/. accessed 4 November 2013.

Xin, L., H. Hong-Yin and Y. Jia. 2010. Lipid accumulation and nutrient removal properties of a newly isolated freshwater microalgae, *Scenedesmus* sp. LX1, growing in secondary effluent. New Biotech. 27(1): 59–63.

Yang, Y.F., CP. Feng, Y. Inamori and T. Maekawa. 2004. Analysis of energy conversion characteristics in liquefaction of algae. Resour. Conserv. Recycl. 43: 21–33.

Zhang, X., Q. Hu, M. Sommerfeld, E. Puruhito and Y. Chen. 2010. Harvesting algal biomass for biofuel using ultra filtration membranes. Biores. Tech. 100: 5297–304.

Zhang, Y., M.A. Dube, D.D. McLean and M. Kates. 2003. Biodiesel production from waste cooking oil: 2. Economic assessment and sensitivity analysis. Biores. Tech. 90: 229–40.

Zheng, S., M. Kates, M.A. Dube and D.D. McLean. 2006. Acid catalyzed production of biodiesel from waste frying oil. Biomass & Bioenerg. 30: 267–72.

Zullaikah, S., C.C. Lai, S.R. Vali and Y.H. Ju. 2005. A two-step acid-catalyzed process for the production of biodiesel from rice bran oil. Biores. Tech. 96(17): 1889–96.

CHAPTER 4

Biorefinery Approach to the Use of Macroalgae as Feedstock for Biofuels

Ana M. López-Contreras,[1,] Paulien F.H. Harmsen,[1] Xiaoru Hou,[2] Wouter J.J. Huijgen,[3] Arlene K. Ditchfield,[5] Bryndis Bjornsdottir,[4] Oluwatosin O. Obata,[5] Gudmundur O. Hreggvidsson,[4,6] Jaap W. van Hal[3] and Anne-Belinda Bjerre[2]*

1. Introduction

Macroalgae (also called seaweeds) have gained attention in recent years as feedstock for the production of fuels and chemicals. This is due to their advantages over traditional terrestrial feedstocks for biorefinery: higher productivity cultivation (amount of biomass produced per unit of surface area) than terrestrial crops, no competition for arable land, lower fresh water consumption during cultivation, and no requirement for fertilizer (van den Burg et al. 2013). In addition, macroalgae have a distinctive chemical composition that differs from lignocelluloses and terrestrial crops, and some

[1] Wageningen Food and Biobased Research (WFBR), Wageningen Research, Bornse Weilanden 9, 6709WG Wageningen, The Netherlands.
[2] Danish Technological Institute (DTI), Gregersensvej, 2630 Taastrup, Denmark.
[3] Energy research Centre of The Netherlands (ECN), Biomass & Energy Efficiency, Westerduinweg 3, 1755 LE, Petten, The Netherlands.
[4] Matis, Vinlandsleid 12, 113 Reykjavik, Iceland.
[5] Scottish Association for Marine Science (SAMS), Scottish Marine Institute, Oban PA37 1QA, United Kingdom.
[6] Faculty of Life Environmental Sciences, University of Iceland, Saemundargata 2, 101 Reykjavik, Iceland.
* Corresponding author: ana.lopez-contreras@wur.nl

species are rich in carbohydrates, proteins, fatty acids, and/or bioactive components that make them very suitable for biorefinery (Kraan 2013, van den Burg et al. 2013). For the production of fuels, the most studied routes are the biological conversion of sugars into liquid fuels such as ethanol or butanol, the thermochemical conversion of macroalgae biomass into liquid fuel by hydrothermal liquefaction (HTL), the chemo-catalytic conversion of sugars into furans, and the anaerobic digestion of biomass into methane. Various reviews of the use of macroalgae for biofuels have appeared in recent years, including Chen et al. 2015, Jiang et al. 2016, Milledge et al. 2014, Suutari et al. 2015, Wei et al. 2013. Chen et al. (2015) concluded that biodiesel production from macroalgae seems less attractive than that from microalgae, given the low content of lipids in macroalgae. Therefore, biodiesel from macroalgal lipids was left outside the scope of this chapter.

This chapter will present an overview of the types of macroalgae and their (chemical) properties. It will also address processes for isolating carbohydrates for biofuel production, and biochemical and thermochemical conversion processes for transforming these carbohydrates into liquid biofuels and biomethane. In addition, it will discuss uses of the proteins in macroalgae as part of the wider picture of using seaweeds as feedstock for the future biobased economy.

2. Macroalgae and Cultivation for Biofuels

Macroalgae are eukaryotic marine multicellular algae that are adapted to live in seas and oceans. They are photoautotrophic, which means they can fix their carbon by photosynthesis (Reece and Campbell 2011). Macroalgae are classified in main groups (called phyla), based on the colour of their photosynthetic pigments (Fig. 4.1). These phyla include Chlorophyta and Streptophyta (green algae), Rhodophyta (red algae), and Phaeophycae (brown algae) (Suutari et al. 2015). Macroalgae belong to several kingdoms, since they are not always evolutionarily closely related.[1] Various macroalgae

Figure 4.1. The three main classes of macroalgae based on their pigment colours. Species illustrated are green algae (*Ulva lactuca*, left), red algae (*Palmaria palmata*, centre), and brown algae (*Laminaria digitata*, right). Pictures are from ©M.D. Guiry,[1] used with permission.

---

[1] http://www.seaweed.ie

have holdfasts, which anchor them to the seabed. A stem-like stripe connects the leaf-like blades to the holdfast. The blades provide the largest surface of the macroalgae thallus and are responsible for most of the photosynthesis (Reece and Campbell 2011).

Seaweed can be obtained by harvesting natural stocks, by collecting drift macroalgae, or by cultivation in aquacultures. In Europe (e.g., Norway, France, Ireland), it is most common to obtain seaweed biomass by harvesting natural stocks in coastal areas with rocky shores and a tidal system. Depending on water temperature, some groups will dominate, like brown seaweeds in cold waters and reds in warmer waters (Bruton et al. 2009). Drift seaweeds are another source. The macroalgae accumulate in lines left behind by the receding tide. The location and seasonal availability of these resources are unpredictable. They have traditionally been collected by coastal communities on a small scale to use as a fertilizer or soil conditioner. Drift *Ulva* are commonly encountered, and they tend to develop at more and more locations along European coasts due to eutrophication (Bruton et al. 2009).

Macroalgae can also be generated through cultivation. Only a dozen algae are commercially cultivated, including *Laminaria, Porphyra, Undaria, Gracilaria, Euchema, Ulva,* and *Chondrus*. Most large cultivation sites for macroalgae are in Asia, and use traditional labour-intensive techniques.[2]

Current macroalgae production is estimated to be 16 million tonnes of wet biomass. China is by far the largest producer (70%), followed by the Philippines, Indonesia, and Korea (together 20%). With 0.15 million wet tonnes, Norway is the largest European producer of seaweed. Seaweed from natural sources (wild harvest) accounts for 5% of the total global seaweed production. Five genera—*Laminaria, Undaria, Porphyra/Pyropia, Euchema,* and *Gracilaria*—represent 76% of the total tonnage of seaweed production by aquaculture (Roesijadi et al. 2010). The amount of cultivated seaweed in the world has continuously increased over the last 10 years at an average of 10% (Fig. 4.2; FAO 2012), particularly due to an increase in red algae production.

A study by the Dutch research institute ECN (Reith et al. 2005) estimated the amount of fuels that could be produced if 10% of the Dutch part of the North Sea were used to cultivate macroalgae. Assuming 50 tonnes of dry matter of seaweed per ha per year, they arrived at 25 million metric tonnes of seaweed (dry matter basis), representing about 6 million metric tonnes of fuels (Kraan 2013, Reith et al. 2005). This would require 5,000 km² of cultivation area. The Crown Estate has indicated that they foresee a cultivation area of 15,000 km² for the UK (Lewis 2011). These numbers indicate that significant amounts of seaweed-based biofuels can be produced

---

[2] The seaweed site: information on marine algae (www.seaweed.ie)

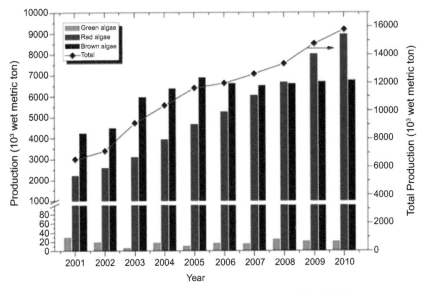

Figure 4.2. Amount of cultivated seaweed in the world (FAO 2012).

if large-scale cultivation is achieved in the North Sea. Other initiatives have been started around the world, for instance in India (SEA6 energy).[3] Finally, Ecofys estimates that seaweed can contribute significantly to greening the fuel supply in the entire world, with a theoretical maximum of 515 EJ or about 34,000 MMT of seaweed or 8,600 MMT of fuel alcohols (Florentinus et al. 2014).

Unlike large-scale cultivation technologies for food crops and lignocellulosic biomass, those for the macroalgae needed for the biofuel industry have not yet been developed to a sufficient scale (Jiang et al. 2016). A recent example of a new, promising large-scale cultivation technology is the use of novel textiles to anchor seaweeds in a high-density manner. This was developed in the European project AtSea[4] and further optimized during the Macrofuels project.[5]

Sustainability in the cultivation of macroalgae is further increased when it is carried out in Integrated Multi-Trophic Aquaculture (IMTA) systems. In IMTA cultivation, the farming of fed (fish, crustaceans) and extractive (macroalgae and shellfish) species is integrated (Barrington et al. 2009; Fig. 4.3). In recent decades, intensive animal aquaculture has faced increasing pressure to minimize its environmental impact. Aquaculture

---

[3] http://www.sea6energy.com/
[4] http://www.atsea.eu
[5] http://www.macrofuels.eu

produces phosphorus and nitrogen in large quantities that are lost to the surrounding ecosystem, and result in eutrophication of the sea environment and economic losses. Over 67–80% of nitrogen and 50% of phosphorus fed to farmed fish goes into the environment, either directly from the fish or from solid wastes.

Macroalgae can use the sun's energy to extract inorganic nutrients (including N and P from fish farming) from the water and to produce new biomass through a process similar to that used by land plants: photosynthesis and reducing eutrophication, which results in the diversification of products from aquaculture and enhances the farms' sustainability and economic viability. This ecological function of algae has been widely documented in scientific publications (Abreu et al. 2011, Fei 2004), and the efficiency of IMTA in addressing the environmental issues associated with intensive marine animal aquaculture has been amply demonstrated on both experimental and pilot scales (Neori et al. 2004, Troell et al. 2003, IDREEM).[6] The combination of efficient large-scale cultivation of macroalgae in IMTA conditions is currently seen as a sustainable way to produce enough biomass for the production of biofuels, but developments of the cultivation systems are still needed.

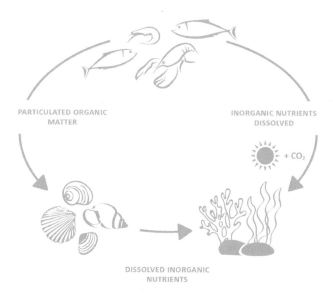

Figure 4.3. Scheme of an IMTA cultivation system for macroalgae and fish (source: Algaplus,[7] used with permission).

[6] http://www.idreem.eu
[7] http://www.algaplus.com

3. Macroalgae Chemical Composition

The chemical composition of macroalgae is significantly different from that of terrestrial plants. In general, seaweeds are characterized by high water content (70–90 wt% fresh weight), high carbohydrate content (brown 30–50 wt% dry weight, red 30–60 wt% dry weight, green 25–50 wt% dry weight), high content of minerals such as alkali metals (10–50 wt% dry weight), low protein content (7–15 wt% dry weight), and low lipid content (1–5 wt% dry weight) (Jung et al. 2013).

Macroalgae show large variations in the content of the different components. This variation depends on the species, as well as on factors such as seasonality, harvesting location, and age of the plant. The content of sulfur, nitrogen, and other minerals varies according to growth stages, and the optimal time for macroalgae harvesting should be selected according to the growth cycle (Ross et al. 2008). For example, Schiener et al. (2015) found that the composition of brown macroalgae varies in different seasons (e.g., highest carbohydrate content in autumn and lowest in winter, which impacts the content of protein, ash, etc.). This can be beneficial for industries that aim to maximize yields of targeted products and minimize less valuable components. For economic and technical feasibility studies, each group of seaweed should be looked at separately, and the optimal cultivation and harvesting conditions need to be determined for the products of interest.

Green macroalgae (e.g., *Ulva lactuca*) have a green colour caused by chlorophyll a and b, which is present in the same proportions as in higher plants. In addition to cellulose, *U. lactuca* contains a significant number of ulvans in their cell walls. These are sulphated polysaccharides which mainly consist of disaccharides that contain sulphated rhamnose linked to glucuronic acid or iduronic acid (Lahaye and Robic 2007). Sometimes, the uronic acids are replaced by xylose or sulphated xylose (Fig. 4.4; Cunha and Grenha 2016).

Red macroalgae have a red colour which is caused by the phycoerythrin and phycocyanin pigments. These mask the green colour of the chlorophyll a, beta-carotene, and a number of unique xanthophyll pigments. Red macroalgae can be divided into three subgroups based on their cell wall polysaccharides. The first subgroup, including *Palmaria palmata*, contains the cell wall polysaccharide xylan (Lahaye et al. 2003). Xylan consists of the repeating pentose sugar xylose (Fig. 4.5). *P. palmata* builds reserves using floridean starch and non-structural soluble floridoside. Floridean starch is a polysaccharide built-up of α-1,4-glucosidic linked glucose. Floridoside is a non-structural glycerol galactoside.

The second red macroalgae subgroup, including the *Kappaphycus* species, is rich in carrageenan as polysaccharide (Baghel et al. 2014, Costa et

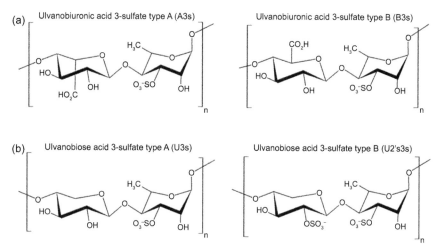

Figure 4.4. Structure of the main repeating disaccharides in ulvan. In green macroalgae, ulvan contains ulvanobiuronic acids (a) consisting of sulphated rhamnose linked to glucuronic acid (A3s) or iduronic acid (B3s). In ulvanobiose acids (b), the uronic acids are replaced by xylose (U3s) or sulphated xylose (U2's3s) (Cunha and Grenha 2016).

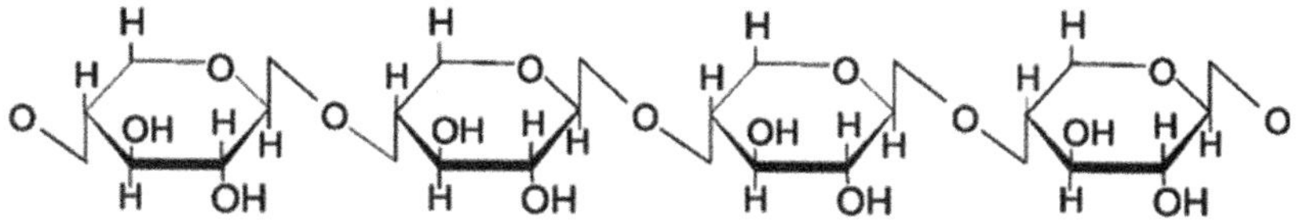

Figure 4.5. Structure of xylan. It is a repeating β-1,4 linked xylose polysaccharide which is present in the cell walls of some red macroalgae.

al. 2012, Villanueva et al. 2010), whereas the third group of red macroalgae, including *Gracilaria* species, produces agar as a structural polymer. The primary structure of agars and carrageenans are similar since they share the repeating disaccharide containing 1,3-linked α-D-galactopyranose and 1,4-linked β-galactopyranose. The only difference between these polysaccharides is that agars contain β-galactopyranose in the L-form, and carrageenans contain β-galactopyranose in the D-form.[8] Carrageenan is mainly used as a gelling agent in the food industry. Various carrageenan structures exist; the kappa (κ), iota (ι), and lambda (λ) forms are the most interesting for the food industry (Cunha and Grenha 2016).

The most interesting species of brown algae for biofuel production are the laminariales order (also known as kelps), with *Laminaria digitata* as the type species. Kelps grow in the intertidal zone or in deeper seawaters,

---

[8] http://www.cybercolloids.net/information/technical-articles/introduction-carrageenan-structure.

and have evolved unique adaptations to resist the crushing of the waves, the drying atmosphere, and the strong solar rays during low tides (Reece and Campbell 2011). They can grow to a length of 60 m, and have strong cell walls consisting of cellulose, alginate, and fucoidan. Alginate is a polysaccharide built of mannuronic acid (M) and guluronic acid (G) (as shown in Fig. 4.6) (Gaserod 2011, Jard et al. 2013). Fucoidan is a sulphated fucose polysaccharide (Pangestuti and Kim 2011). Alginate and fucoidan are widely used in the food and pharmaceutical industries.

Kelps store their reserves in the form of the complex polysaccharide laminarin and the non-structural sugar alcohol mannitol. Laminarin is a polymer consisting of β-1,3-linked glucose monomers, with ramifications of β-1,6-linked glucose monomers (Fig. 4.7) (Graiff et al. 2016). Laminarin is the most important reserve polysaccharide in laminariales, accounting

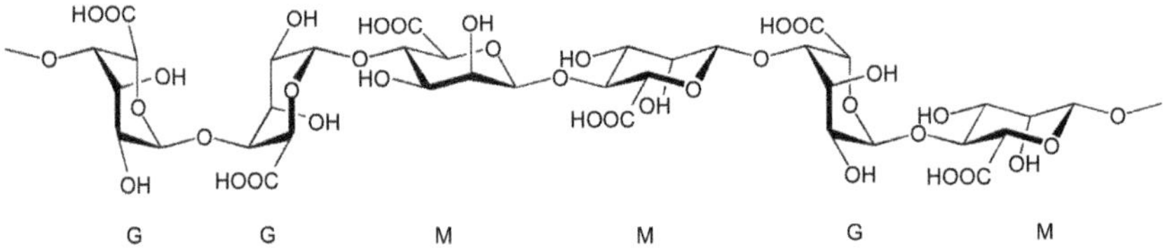

Figure 4.6. Structure of alginate. Alginate is composed of mannuronic acid (M-blocks) and guluronic acid (G-blocks). The structure of alginate depends on the distribution between repeating M, G, or MG blocks (Gaserod 2011, Jard et al. 2013).

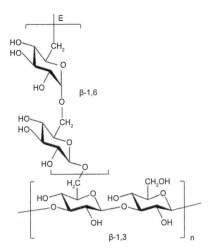

Figure 4.7. Structure of laminarin. Laminarin is a polymer consisting of β-1,3-linked glucose monomers, with ramifications of β-1,6-linked glucose monomers.

for up to 50% of the dry matter in *Laminaria digitata,* and 47% of the dry matter of *Saccharina latissima* when cultivated at low salinity conditions on the Danish coast (Nielsen et al. 2016). In the same Danish study, the mannitol content of *L. digitata* varied between 7 and 21% of the dry matter, and between 12 and 23% of the dry matter for *Saccharina latissima;* these are the highest concentrations determined during cultivation in low salinity waters. Both laminarin (after hydrolysis to soluble glucose) and mannitol represent good substrates for fermentation into biofuels.

4. Cascading Biorefinery Approach to Macroalgae

The three main components in macroalgae are carbohydrates, proteins, and minerals. In contrast to microalgae, macroalgae are not rich in lipids. Macroalgae contain a vast array of high-value bioactive compounds that find applications in pharmaceuticals, cosmetics, health foods, and natural pigments. The main applications of macroalgal products nowadays are as food, as feedstock for the hydrocolloids alginate, carrageenan, and agar from brown and red species, and as feed additives and fertilizer.

The economic feasibility of seaweed biofuel production would be significantly enhanced by a high-value co-product strategy. According to Suganya et al. (2016), this strategy would involve cultivating algae with an algal farming facility (CO_2 mitigation, wastewater treatment), extracting bioactive products from the algae, applying thermal processes (pyrolysis, liquefaction, gasification) to the extracted biomass, extracting high-value chemicals from the resulting liquid, vapour, and/or solid phases, and reforming/upgrading biofuels for different applications. Brennan and Owende (2010) discussed the technologies underpinning microalgae-to-biofuel systems, focussing on biomass production, harvesting, conversion technologies, and the extraction for useful co-products. This chapter discusses cascading biorefinery approaches for macroalgae, focussing on the carbohydrates for biofuel production and on the protein fraction.

Cascading biorefinery approaches aim to use the maximum inherent value of all components present in the biomass, as an alternative to single-product approaches. In general, the cascading approach for seaweed is based on fractionating the carbohydrate molecules, proteins, and minerals, separating the different carbohydrate molecules from each other, and converting them into high-value chemical intermediates, or using them as is. Macroalgae are ideally suited for this approach because they contain both high-value components and carbohydrates for biofuel production (van Hal et al. 2014).

Pre-treatment technologies for lignocellulosic biomasses often require harsh conditions (using high temperature, high pressure, and/or the addition of chemicals) to open the cellulosic and hemicellulosic polymers for subsequent enzymatic hydrolysis of sugars to fermentation. This requires costly equipment and generates compounds that inhibit fermentation due to degradation products formed from sugars and lignin (Klinke et al. 2003, Palmqvist and Hahn-Hägerdal 2000). These compounds negatively affect the fermentation processes (Sassner et al. 2008). Due to the lower crystalline structures of the sugar polymers and the lack of lignin in macroalgae, mild pre-treatments are sufficient to realize high yields of sugars solubilisation by enzymatic hydrolysis of sugar polymers. It has been reported that minor pre-treatment in the form of size reduction (e.g., by milling or cutting the macroalgae leaves) is sufficient for facilitating enzymatic hydrolysis for green macroalgae (Schultz-Jensen et al. 2013), even by means of commercial enzymes developed for 2G feedstocks for brown seaweed (Hou et al. 2015). Section 5 will give an overview of different pre-treatments used for different macroalgal types.

4.1 Carbohydrates for Biofuel Production

Carbohydrates are the feedstock for the production of biofuels by fermentation. In macroalgae, the carbohydrates can be divided into storage carbohydrates, which function as a food reserve, and structural carbohydrates, which are found in the cell wall and give mechanical strength and prevent the seaweed from dehydration. Some of these carbohydrates are also found in terrestrial plants, while others are exclusively found in seaweeds.

Table 4.1 summarizes the carbohydrates present in brown, red, and green seaweeds; the carbohydrates in bold are suitable feedstocks for fermentation. In general, these are neutral sugars with limited amounts of sulphated groups. Negatively charged carbohydrates, such as alginates, are best suited for other applications. For the production of biofuel from seaweed, it is wise to select species and harvest periods with an optimal content of neutral carbohydrates.

Most studies on macroalgal biorefineries focus on brown species like *Saccharina, Laminaria,* and *Sargassum.* This is probably because these species have been traditionally consumed, and thus extensively cultivated and researched in East Asian countries (Jiang et al. 2016). Research has focused on utilizing all possible carbohydrates in fermentation (e.g., co-utilization of glucose, mannitol, and uronic acids). Programmes looking at the seaweed biorefinery approach, like the Dutch Seaweed Biorefinery project[9] and the Danish MAB3 project,[10] have also considered products besides biofuels to

---

[9] http://seaweed.biorefinery.nl
[10] http://www.mab3.dk

Table 4.1. Carbohydrates present in brown, red, and green macroalgae, adapted from Wei et al. (2013).

Macroalgae % d.m. carbohydrates	Storage carbohydrates		Structural carbohydrates	
	Carbohydrate	*Building block*	*Carbohydrate*	*Building block*
Brown 30–50 wt%	Laminarin	**Glucose**	Cellulose	**Glucose**
	Mannitol	**Mannitol**	Alginate	**Uronic acids** (guluronic acid, mannuronic acid)
			Fucoidan	Fucose (sulphated)
Red 30–60 wt%	Floridean starch	**Glucose**	Cellulose	**Glucose**
			Xylan	**Xylose**
			Carrageenan	**Galactose, 3,6-anhydrogalactose** (sulphated)
			Agar	**Galactose** (scarce sulphated)
Green 25–50 wt%	Starch	**Glucose**	Cellulose	**Glucose**
			Ulvan	**Rhamnose, xylose, glucuronic acid** (sulphated)

improve the process economics of future biorefineries by co-production of high-value products. Brown seaweeds are high in valuable alginate, a hydrocolloid for food and pharmaceutical applications. The building blocks from alginate with acid functionalities might also be converted to furan dicarboxylic acid or similar acids, as building blocks for polyesters (López-Contreras et al. 2014, van Hal et al. 2014).

For the red macroalgae, species rich in xylan like *Palmaria palmata* are very suitable feedstocks for biofuel production. Mutripah et al. (2014) evaluated 20 species of seaweed as resources for ethanol production. Additional method refinement was performed with *Palmaria palmata*, the most promising candidate. Various parameters of acid hydrolysis (catalyst amount, hydrolysis temperature, and time) and fermentation (inoculum concentration, fermentation period, and medium pH) were then adjusted to obtain the highest possible ethanol yield. Other red seaweed types rich in agar and carrageenan find their way in the hydrocolloid industry rather than in the biofuel industry, due to the high prices of red-algae-derived hydrocolloids. For these species, the side streams from hydrocolloid extraction could still be utilized for biofuels.

Green macroalgae are attracting interest as a carbohydrate resource due to the comparatively high level of accessible sugars, especially starch and cellulose. Green seaweeds have high water content and are very sensitive to microbial degradation, generate H_2S, and need to be processed shortly after harvest. Recently, Bikker et al. (2016) reported an example of an *Ulva lactuca*-based biorefinery in which the carbohydrates were fermented to acetone, butanol, ethanol, and 1,2 propanediol, and the protein-rich fraction was used as animal feed.

Figure 4.8 shows the conceptual process scheme from the Dutch TO2 project for processing *Ulva* sp. amongst others to biofuels (i.e., acetone, butanol, and ethanol, ABE; Groenendijk et al. 2016). After size reduction, the first step is to separate the proteins in the biomass. The sugar fraction is then hydrolysed by mineral acid (here, hydrochloric acid is shown as an example, but sulfuric acid might be an alternative). After neutralization, the rhamnose is removed by a simulated moving bed reactor. The remaining organics are then used to produce ABE by fermentation. ABE is concentrated by distillation and the other fermentation products are converted into fertilizer feedstock by reverse osmosis. The major uncertainty in this process is the feasibility of protein removal.

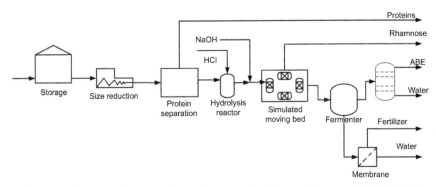

Figure 4.8. Conceptual process scheme for processing *Ulva* sp. (Groenendijk et al. 2016).

4.2 Proteins

Intact seaweed and seaweed fractions are potential novel protein sources for animal nutrition (Holdt and Kraan 2011). The protein content of macroalgae not only differs between species, but it also largely depends on the seasonal harvesting period and the cultivation conditions. In general, brown species have a lower protein content than green and red species. Among species available in European temperate Atlantic waters, *Ulva* sp. have been extensively characterized (Fleurence 1999) and exhibit a high crude protein content (up to 44% of dry matter; Holdt and Kraan 2011). For the red macroalgae *Palmaria palmata*, protein contents between 9% dry matter during the summer months and 25 wt% dry matter by the end of the winter and spring periods have been reported (Fleurence 1999).

Protein is generally known to be the most expensive dietary source for fish feeding. The traditional source is fishmeal, which represents over 50% of operating costs in intensive aquaculture (Lovell 2003). There have been many efforts to find cheap alternative protein sources (e.g., plant protein from soybeans, peas, and seaweeds; Nunes et al. 2014). Several studies have described the replacement of proteins for aquaculture by macroalgae. By adding *Gracilaria bursa-pastoris* and *Ulva rigida* (up to 10% dry matter), or *Gracilaria cornea* (up to 5% dry matter) in fish feed for sea bass, Valente et al. (2006) found effects that were comparable to those of traditional fish feed on fish growth performance and feed utilization efficiency. They thus claimed that seaweed proteins could potentially be used as an alternative/complementary protein source for fish feed.

Proteins are part of the cell wall and are closely associated with carbohydrates. Extraction of proteins from seaweed is difficult due to the presence of these carbohydrates, as they increase viscosity and limit access to the proteins. In the Dutch TO2 project (Groenendijk et al. 2016), two lines were followed for protein extraction (i.e., alkaline extraction followed by isoelectric precipitation, and carbohydrate/ulvan hydrolysis for the production of protein-enriched fractions). To summarize, alkaline extraction followed by isoelectric precipitation was not successful, as proteins were poorly released from the seaweed biomass. The production of protein-enriched fractions by hydrolysis of carbohydrates was more successful, as sugar release was observed. Large differences were detected between freeze-dried seaweed and fresh biomass; the release of sugars from fresh seaweed is much higher than from freeze-dried material. Membrane filtration proved to be an interesting technique for the desalination and purification of samples. Further work needs to focus on cell disruption of seaweed, comparable to work done on microalgae. Microscopic analysis found that most of the *Ulva* cells were still intact even after washing, homogenisation, and enzymatic hydrolysis over a few days.

Hou et al. (2015) suggested a novel integrated process for lifting the algal protein content in fractions of *Laminaria digitata* (due to the hydrolysis of polymers in the cell wall matrix) as a high-value by-product, along with ethanol production. A recent study on the biorefinery of *Ulva lactuca* for the use of carbohydrates for fermentation to ABE and 1,2-propanediol found that the *Ulva lactuca* extracted fraction after the solubilization of sugars for fermentation is enriched in total protein (34% of the dry matter compared to 22% of the dry matter in the initial biomass). This extracted fraction seems to be a promising protein source in diets for monogastric animals, with better characteristics than the intact *U. lactuca*. *In vitro* gas production tests indicated a moderate rumen fermentation of *U. lactuca* and the extracted fraction, similar to that of alfalfa (Bikker et al. 2016).

5. Isolation and Extraction of Sugars from Macroalgae

5.1 Handling

Fresh seaweed has a high water content: 70–90% of its fresh weight is water. For further processing, the seaweed needs to be dewatered or dried after harvesting to reduce the weight of the biomass and reduce transportation volume and costs. Dewatering can improve shelf life and improve seaweed handling (Bruton et al. 2009). However, drying may increase the costs of the overall process, as energy is required to evaporate the water. In addition, some processes require wet biomass (e.g., fermentation processes), and drying can be avoided if the seaweed biomass can be stabilized prior to processing.

Fresh seaweed often also contains foreign objects like stones, snails, or other undesired material that might need to be removed before processing. Manually harvested seaweeds often contain less debris, while drift seaweeds often contain a large amount of sand that needs to be removed (Bruton et al. 2009).

For some applications, such as certain fermentation processes, samples must contain a low level of salt. Desalination of seaweed requires washing with fresh water and is costly, and downstream processes should be adapted to avoid a desalination step where possible. This section describes the extraction of sugars for biofuel production for each type of seaweed, considering the differences in seaweed composition.

5.2 Examples of Physical, Chemical, and Enzymatic Approaches to Yield Sugars

Mannitol has been isolated from *Saccharina latissima* using a simple pressing step. Pressing fresh seaweed biomass in a screw press results in a press cake (70–75 wt%) and a press liquid (25–30 wt%) (López-Contreras et al. 2014). The increase of the dry matter content of the press cake was limited, but the total sugar content of the press cake was reduced with 20 wt%, mainly due to the removal of mannitol. Analysis of the press liquid revealed that the mannitol concentration was 16 g/L.

Chemical hydrolysis often involves a combination of acids and high temperatures to disrupt macromolecular structures. The optimal type of acid, concentration, temperature, and treatment duration depends on the algal species. Though this can be a very effective method to disrupt macroalgal biomass, it lacks specificity and is often detrimental to protein nativity and solubility, decreasing the potential value of the protein fraction.

Wang et al. (2011) developed a saccharification method to produce ethanol from *Gracilaria*. Dilute acid hydrolysis of the homogenized seaweed biomass yielded low concentrations of glucose (4.3 g glucose/kg fresh biomass), whereas a two-stage hydrolysis (combination of dilute acid hydrolysis with enzymatic hydrolysis) produced 13.8 g glucose/kg fresh biomass (Wang et al. 2011).

Palmaria palmata is an edible red species that has been consumed for centuries for its proteins (Galland-Irmouli et al. 1999, Morgan et al. 1980), dietary fibres (Lahaye et al. 1993), vitamins (A & C), and minerals (iron, magnesium, calcium, and iodine) (Morgan et al. 1980). Its main sugar polymer is xylan, which can contribute up to 35% of the dry weight (Kraan 2012). Production of carbohydrates from the species *Palmaria palmata* has been studied using catalysts such as xylanase enzymes (Lahaye and Vigouroux 1992).

For *Ulva lactuca*, van der Wal et al. (2013) reported the solubilisation of 75–90% of the total carbohydrates by applying water extraction at 150°C or at 85°C for 10 minutes, followed by enzymatic hydrolysis using commercial cellulases. Pre-treatments using diluted acid or alkali under the same conditions did not improve yields of sugar solubilisation (van der Wal et al. 2013). In another study, acid hydrolysis at an elevated temperature (120–150°C) was applied to solubilize sugars without enzymatic hydrolysis for *Ulva* biomass. The expected advantages of this process include lower costs, shorter hydrolysis time, and simple operation (Jang et al. 2012).

Enzymatic hydrolysis is used to solubilize fermentable sugars from the sugar polymers in biomass, in principle by cleaving the bonds between the monomers with the addition of water molecules. The reactions between enzymes and substrate (i.e., sugar polymers in this context) are often referred to as 'lock-key' reactions. Different enzymes are required for hydrolyzing different substrates at their specific reaction sites (e.g., laminarinase (β-glucanase) catalyzes the hydrolysis of polysaccharides β-glucans into glucose, while xylanases catalyze the hydrolysis of polysaccharides β-xylans into xylose). At the moment, there are no specific enzymatic cocktails commercially available for macroalgae degradation. Therefore, the cocktails developed for lignocellulosic biomass are being used.

As an example of the enzymatic hydrolysis for fermentation, Hou et al. (2015) reported the enzymatic hydrolysis of a minor pre-treated (i.e., only by drying and milling) brown seaweed *L. digitata*. This brown macroalgae showed a high glucose content (i.e., 56% in dried weight), which was present mainly in the form of laminarin. As was introduced in Section 3, laminarin is a polysaccharide composed of a backbone of β-1,3-glucan with ramifications connected through β-1,6-glucan bonds. The hydrolysation of such a glucan matrix into monomer glucose requires synergistic work from at least three groups of enzymes: β-1,3-glucanases, β-1,6-glucanases, and β-glucosidase. β-1,3-glucanases and β-1,6-glucanases hydrolyse β-1,3-glucans and β-1,6-glucans to glucose and disaccharides, and β-glucosidases hydrolyse the disaccharides into glucose. Figure 4.9 shows a simplified model of this synergistic hydrolysis of laminarin by β-glucanases and β-glucosidase.

Dried and milled *L. digitata* biomass has been subjected to enzymatic hydrolysis by adding a determined amount of water and the commercial enzyme mixture Celluclast 1.5L. Celluclast 1.5L contains high β-glucanase, β-glucosidase, xylanase, and β-xylosidase activities (Thygesen et al. 2003). Different loadings of Celluclast 1.5L were tested for the hydrolysis of the biomass. The results showed an obvious positive correlation between enzyme loadings and glucose recovered (Fig. 4.10). It is worth mentioning that when enzyme loadings reach saturation points (e.g., reaction sites between enzymes and substrates are all occupied), an increase in enzyme loadings will not further increase the amount of solubilized glucose.

Moreover, Hou et al. (2015) studied alginate degradation in the biomass by adding alginate lyase to Celluclast 1.5L. The addition of the lyase enzyme

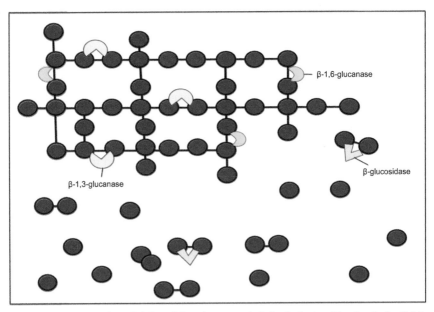

Figure 4.9. Simplified model describing the synergistic hydrolysis of laminarin by β-1,3-glucanases, β-1,6-glucanases, and β-glucosidases.

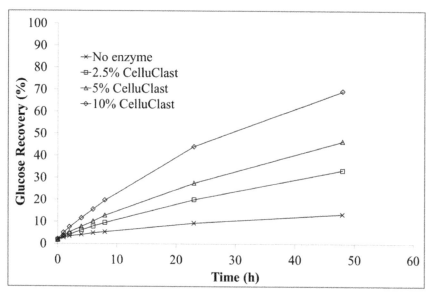

Figure 4.10. Time course of glucose release during enzymatic hydrolysis of *L. digitata*, by Celluclast 1.5L at different loadings (Hou et al. 2015).

improved the glucose recovery slightly, from 69% (enzyme loading: 10% v/w Celluclast 1.5L) to 77% (enzyme loading: 10% v/w Celluclast 1.5L + 0.12% w/w alginate lyase) to 80% (enzyme loading: 10% v/w Celluclast 1.5L + 0.5% w/w alginate lyase). Alginate lyases degrade alginate through β-elimination of the glycosidic bonds between mannuronic acids and guluronic acids (Kim et al. 2011). This degrades the alginate for fermentation by uronate utilizing organisms, and helps to remove the structural blockage for enzymes contacting the glucans that are trapped in the alginate matrix.

The use of thermostable pre-processing enzymes might be particularly important and interesting for the processing and hydrolysis of viscous macroalgal biomass. High temperatures increase the solubility of polysaccharides and lead to reduced viscosity, enabling higher feedstock loads and facilitating enzymatic access to polysaccharides. Furthermore, thermostable carbohydrate-processing enzymes that originate from thermophiles found in extreme environments are inherently robust and suitable for a range of biorefinery applications (Turner et al. 2007). Thermostable macroalgal processing enzymes are not yet available, but efficient alginate lyases have been expressed in high yields and are currently under development (Hreggvidsson et al. 2015).

6. Thermochemical Conversions to Fuels

6.1 Hydrothermal Liquefaction

The thermochemical processing of whole macroalgae to produce fuels can either be performed by dry pyrolysis, or in water slurry by hydrothermal liquefaction (HTL) (Elliott 2016). Given the high water content of seaweeds, using HTL to liquefy seaweed and produce a bio-crude seems like a logical strategy. Hydrothermal treatment can be performed at either sub- or supercritical conditions. For example, Anastasakis and Ross (2011) reported a bio-crude yield of 19.3% dry weight from *Saccharia latissima* (formerly *Laminaria saccharina*) using a treatment at 350°C during 15 min without a catalyst in a batch reactor. The bio-crude phase had a higher heating value (HHV) of 36.5 MJ/kg. Alkaline earth metals (e.g., Ca and Mg) predominantly ended up in a solid residue, while alkaline metals (e.g., Na and K) ended up in the aqueous phase.

Anastasakis and Ross (2015) used the same approach in a follow-up study that also included other brown macroalgae (i.e., *L. digitata, L. hyperborea*, and *A. esculenta*). They reported bio-crude yields from 9.8 to 17.8 wt% with a HHV of 15.7–26.2 MJ/kg. López-Barreiro et al. (2015) reported a somewhat higher bio-oil yield of 29.4 wt% from *A. esculenta* at 360°C using batch micro autoclaves. In addition to kelps, HTL can also be applied to other macroalgae types. For example, Neveux et al. (2014) applied HTL to various marine and freshwater green macroalgae including *Oedogonium, Derbesia*, and *Ulva*. This resulted in bio-crude yields of up to 26.2 wt% and 19.7 wt% for freshwater and marine species, respectively.

Research by NTNU and SINTEF in Norway found that a high heating rate could substantially improve the yield of bio-oil. A bio-oil yield as high as 79 wt% (dry and ash free) was obtained from *L. saccharina* using the same temperature and residence time, but using sealed quartz capillary reactors and a heating rate of 585°C/min (Bach et al. 2014). Optionally, organic co-solvents can be applied during HTL of macroalgae to fractionate the bio-oil *in situ* (He et al. 2016). Research using the filamentous freshwater macroalgae *Oedogonium* found that the use of a non-polar co-solvent, such as n-heptane, resulted in a bio-crude with improved properties for subsequent hydrotreatment. Hydrothermal liquefaction of macroalgae can also be performed continuously, as demonstrated by Elliott et al. (2013). At a reaction temperature of 350°C, 58% of the carbon was converted into a bio-oil that could be separated by gravity without the use of an additional organic solvent.

The benefits of direct thermochemical conversion of whole macroalgae into a bio-crude followed by hydrotreatment include the production of a hydrocarbon fuel that is compatible with the current infrastructure (Elliott

2016). In addition, these approaches are less dependent on the type and composition of macroalgae, creating the possibility to process various algae feedstocks over a longer period of the year. However, the use of a high temperature and pressure in combination with salts present puts constraints on the equipment that can be used on an industrial scale, especially when supercritical conditions are applied. In addition, López-Barreiro et al. (2015) concluded that the use of HTL to directly produce fuels from macroalgae is not attractive due to the low yield of bio-crude oil in relation to the value of fuels. HTL of the organic residue after extracting valuable components from macroalgae in a biorefinery scheme seems a more promising approach.

6.2 Production of Furans from Macroalgae

An alternative to thermochemical conversion of whole macroalgae is the isolation of the carbohydrate fraction as part of a biorefinery scheme and its subsequent conversion into a fuel (precursor). Thermochemical dehydration of carbohydrates yields organic compounds containing a furan-ring: a five-membered aromatic ring with four carbon atoms and a single oxygen atom. Furans offer interesting properties for application as a biofuel (additive) either directly or after derivatisation (e.g., furfural; Lange et al. 2012). In addition, furans are considered to be platform chemicals from which a large variety of biobased chemicals can be produced (Van Putten et al. 2013). Well-known examples of furans are hydroxymethylfurfural (HMF), which results from dehydrating hexoses such as glucose (Van Putten et al. 2013), and furfural, which results from dehydrating pentoses such as xylose (Zeitsch 2010).

Given the wide variety of carbohydrates present in seaweeds (see Section 4), various corresponding furans could potentially be produced. Examples of possible conversions include:

- Production of furfural from either xylose from red seaweeds like *Palmaria palmata* or alginic acid (uronic acids, brown seaweeds).

- Conversion of glucose from laminarin (brown seaweeds) and/or cellulose into hydroxymethylfurfural (HMF).

- Conversion of rhamnose (from green seaweeds like *Ulva* sp.) into 5-methyl furfural.

The thermochemical conversion process can be performed in water or organic media, or can be solvent-free. Conversion of the seaweed-specific polymeric carbohydrate alginate into furfural has been studied using catalytic hydrothermal approaches (Jeon et al. 2015, 2016), an acid-catalysed process in THF/water (Park et al. 2016), and analytical pyrolysis (Anastasakis et al. 2011, Ross et al. 2009). The latter has been reported to yield furfural with a high selectivity. Conversion of non-seaweed-specific carbohydrates such as xylose and glucose has been extensively discussed in recent reviews (Mariscal et al. 2016, Van Putten et al. 2013).

7. Biological Conversions to Liquid Fuels

7.1 Separate Hydrolysis and Fermentation, and Simultaneous Saccharification and Fermentation Approaches

As described in the previous sections, macroalgae do not contain lignin, and their structural sugar polymers are less crystalline than those in lignocelluloses. Therefore, mild pre-treatments can be applied to efficiently solubilize sugars for fermentation with the subsequent formation of lower amounts of potential fermentation inhibitors (i.e., furfural, lignin components) in the pre-treated streams compared to lignocellulosic hydrolysates. These features make macroalgae interesting feedstocks for simultaneous saccharification and fermentation processes (SSF), in contrast with the most currently used process for lignocellulosic biomasses, which consists of separate hydrolysis and fermentation (SHF) (Fig. 4.11).

SSF was first described by Takagi and colleagues in 1977, in a study that combined the enzymatic hydrolysis of starch with simultaneous fermentation of the sugars to ethanol in the same bioreactor under the same conditions. Such a process has the advantages of reducing the possible product inhibition to enzymes by the accumulation of glucose, lowering

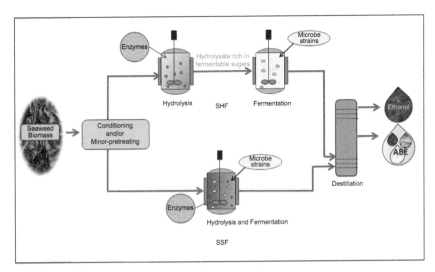

Figure 4.11. Principles of biofuel (ethanol or ABE) production from macroalgae biomass by SSF and SHF.

the capital costs (by using fewer reactors), simplifying the process, and decreasing the risk of contamination (Rudolf et al. 2005). Details of the numerous efforts that have been made to improve ethanol yield by the SSF process can be found in the literature (Philippidis et al. 1993, Wright et al. 1988).

One of the most important requirements for a successful SSF process is the compatibility or compromise of the saccharification and fermentation system, with respect to temperature, pH, mixing rate, substrate concentration, and production of potential inhibitors for the fermentation or hydrolysis. For example, the optimum temperature for enzymatic hydrolysis is typically higher than that of fermentation, at least when using mesophilic yeasts (Olofsson et al. 2008). The optimum temperature for standard cellulases is 45–100°C, while that for ethanol fermentation using *S. cerevisiae* is approx. 30°C. Because of this difference in optimal temperature, a compromise must be made for SSF processes. In such a context, the use of thermotolerant microorganisms has advantages for the SSF process, and many reports on thermophilic ethanol production by thermophilic yeasts or bacteria are available (Olofsson et al. 2008, Scully and Orlygsson 2015, Shahsavarani et al. 2013).

Consolidated bioprocessing (CBP) is a special type of SSF process in which the hydrolysis of the sugar polymers and the fermentation are performed by the same organism, without the addition of external enzymes. Studies of CBP approaches for ethanol production from lignocellulosic biomasses have been reported using thermophilic bacteria, such as *Clostridium thermocellum* (Tian et al. 2016) or *Caldicellulosiruptor bescii* (Chung et al. 2014), and also mesophilic bacteria, such as *Clostridium cellulovorans* for ethanol and butanol production (Yang et al. 2015). For the use of macroalgae, work on the direct fermentation of brown species to ethanol using a CBP strategy has been reported and mainly focuses on the utilization of laminarin and alginates in the biomass (Enquist-Newman et al. 2014, Wargacki et al. 2012).

In SHF, enzymatic hydrolysis and fermentation run separately under each step's optimal condition. The advantage of SHF is that conditions (e.g., temperature, pH, mixing rate for the enzymatic hydrolysis) and fermentation can be optimized independently from each other. The drawback to this process is that substrate inhibition might occur at too high a sugar concentration during enzymatic hydrolysis. A study by Hou et al. (2015) found that the SHF process resulted in a much higher ethanol production yield than the SSF process when using the brown seaweed *Laminaria digitata* as a substrate, mainly due to the higher enzymatic hydrolysis efficiency in the separated hydrolysis step.

Alginate fermentation has been a major challenge. The metabolic engineering of *Escherichia coli* and *Saccharomyces cerevisiae* for production of ethanol from alginate is a ground-breaking accomplishment in the field (Enquist-Newman et al. 2014, Wargacki et al. 2012). Recently, a thermophilic bacterium has been shown to simultaneously utilize glucose, mannitol, and alginate for consolidated ethanol production (Ji et al. 2016). Since alginate constitutes a major fraction of the carbohydrates of brown algae, it needs to be utilized to maximize ethanol yields and improve economic feasibility.

7.2 Ethanol Fermentation

More than 35 billion tonnes of bioethanol is currently produced per year, mainly from first-generation biomass feedstocks (i.e., sugar cane and maize) in Brazil and the USA (FO Licht 2010). Ethanol can also serve as an important chemical building block with special importance for the production of ethylene. Ethylene is the raw material used in the manufacture of the most common polymers such as polyethylene (PE), polyethylene tetra-phthalate (PET), polyvinyl chloride (PVC), and polystyrene (PS), as well as fibres and other organic chemicals (Harmsen et al. 2014).

Baker's yeast (*Saccharomyces cerevisiae*) is the most used ethanol producer and has been used for wine-making, brewing, and baking since ancient times. Baker's yeast is still considered to be the most suitable microorganism for bioethanol production due to its merits: high ethanol tolerance, high osmotic tolerance, and resistance to infections because of the requirement of acidic pH (around 5) for growth. Natural strains of *S. cerevisiae* can ferment C6 sugars (glucose, mannose, galactose), but not C5 sugars (van Maris et al. 2006).

C5 l sugars (xylose, arabinose) and the deoxy sugar (rhamnose) can be fermented by other microbes (as described in Table 4.2). An approach to convert all algal sugars into ethanol could be the sequential fermentation of hydrolysates by different microorganisms. For example, xylose can be fermented into ethanol by another mesophilic yeast strain *Spathaspora passalidarum* (Hou 2012) and/or *Thermoanaerobacter* sp. A *Thermoanaerobacter* J1 strain isolated from a hot spring in Iceland can ferment a wide spectrum of sugars into ethanol (including arabinose and rhamnose).

The final ethanol concentration after fermentation is a critical factor in industrial production of ethanol, due to its effect on the cost of distillation for final ethanol recovery. Several studies have suggested that the final generated ethanol concentration should reach approximately 4–5% to make the process economically feasible (Lynd 1996). In recent years, researchers have investigated the potential to use various seaweed species as substrate for ethanol production (Table 4.2): these include brown seaweed species such as *Undaria pinnatifida* (Cho et al. 2013), *Laminaria digitata* (Hou et al. 2015), and *Saccharina japonica* (formerly *Laminaria japonica*) (Lee et al. 2013); red seaweed species such as *Gelidium elegans* (Yanagisawa et al. 2011) and

Table 4.2. Examples of utilization of seaweeds as feedstock for ethanol and for ABE fermentation.

Macroalgae	Fermentation substrate[1]	Microorganism	Fermentation product	Reference
Red macroalgae				
Gracilaria salicornia	Glucose, xylose	*E. coli*	Ethanol	(Wang et al. 2011)
Kappaphycus alvarezii	Glucose, galactose	*Saccharomyces* sp.	Ethanol	(Hargreaves et al. 2013)
Gracilaria spp.	Glucose, xylose	*Spathaspora pasalidarium*	Ethanol	(Takagi et al. 2015)
Brown macroalgae				
Saccharina latissima	Glucose	*S. cerevisiae*	Ethanol	(Adams et al. 2009)
Laminaria hyperborea	Glucose, mannitol	*Zymobacter palmae*	Ethanol	(Horn et al. 2000a,b)
Laminaria digitata	Glucose	*S. cerevisiae*	Ethanol	(Hou et al. 2015)
Saccharina sp.	Glucose, mannitol	*C. acetobutylicum*	ABE	(Huesemann et al. 2012)
Green macroalgae				
Ulva rigida	Glucose (SSF)	*S. cerevisiae*	Ethanol, glycerol	(Korzen et al. 2015)
Ulva lactuca	Rhamnose, glucose	*C. acetobutylicum, C. beijerinckii*	ABE, 1,2-propanediol	(van der Wal et al. 2013)
Ulva lactuca	Glucose	*C. beijerinckii*	ABE	(Potts et al. 2012)

[1] Refers to the sugars or substrates utilized for fermentation

Gracilariopsis longissima (formerly *Gracilaria verrucosa*) (Kumar et al. 2013); and green seaweed species such as *Ulva australis* (formerly *Ulva pertusa*) (Yanagisawa et al. 2011). However, to our knowledge, only the species *Gelidium elegans Kuetzing*, when subjected to successive saccharification and fermentation, has been documented as exceeding the 4–5% ethanol threshold (Yanagisawa et al. 2011).

The use of thermophilic organisms for a variety of bioconversion processes is of considerable interest (Turner et al. 2007). Thermophiles are robust by nature, living in the harsh, high-temperature environments of geothermal habitats. They are adapted to high-temperature conditions and to the presence of toxic sulfuric compounds and poisonous metal ions and complexes, conditions that reign in high-density raw biomass slurries fed to bioreactors. Fermentation at elevated temperatures reduces the costs of cooling and prevents contamination of spoilage bacteria. It also eases the extraction of many volatile products, either by distillation or gas stripping. This alleviates the potential problem of product inhibition or intolerance, and should prolong the fermenting life of cultures. Furthermore, compared to traditional fermentation organisms, many fermentative thermophiles have the required wide substrate range for utilization of complex carbohydrates, as they produce a great diversity of polysaccharide degrading enzymes, and are potentially capable of simultaneous saccharification and bioconversion using pentose and hexose sugars, including mannitol and even alginate (Chang and Yao 2011, Ji et al. 2016).

An example of a thermophilic organism that can be used to develop microbial bioconversion of macroalgal biomass into ethanol is the anaerobic *Thermoanaerobacterium* sp. strain AK17, which originates from a hot spring in Iceland. The bacterium is easily cultivated and exhibits diverse metabolic activities and efficient fermentation capacities. It has a broad substrate range, efficiently producing ethanol on a variety of hexoses and pentoses, mannitol, and glucuronic acid (Almarsdottir et al. 2012, Koskinen et al. 2008, Sveinsdottir et al. 2009).

7.3 Acetone, Butanol, and Ethanol

Some species of the anaerobic bacterial genus *Clostridium* can produce a mixture of ABE by fermentation from a wide variety of biomasses. This process is known as ABE fermentation. It is now being commercially re-introduced for the production of biologically derived butanol (biobutanol) to be used as biofuel, or to replace petrochemically produced butanol in the bulk chemicals market (Green 2011, López-Contreras et al. 2010).

ABE-producing Clostridial species can utilize all the sugars in plant biomass (López-Contreras et al. 2010), both C5 and C6. In addition, these organisms could be genetically modified to develop strains that can degrade and ferment macroalgal polymers in an SSF or CBP process. An example of this approach is the cloning of a fungal endoglucanase into the ABE producer *C. beijerinckii*, resulting in a strain that can use lichenan (a ramified glucose polymer similar to laminarin) as the sole carbon source for fermentation (López-Contreras et al. 2001). However, the fermentation of macroalgae biomass to ABE using an SSF approach has not yet been reported.

The fermentation of hydrolysates from brown macroalgae to ABE has been reported for several species and microorganisms (Table 4.2). The utilization of mannitol, glucose, and laminarin in extracts from *Saccharina* sp. by *Clostridium acetobutylicum* has been characterized by Huesemann and colleagues (2012). The strain showed a diauxic growth pattern when grown from glucose/mannitol mixtures, with a strong preference for glucose over mannitol. The yields of solvents obtained were somewhat lower than those on control media, partially due to an incomplete acid re-assimilation during fermentation. Interestingly, *C. acetobutylicum* appeared to utilize laminarin in the extract without external addition of enzymes.

A pilot-scale test on the production of butanol from *Ulva lactuca* collected in a bay in New York showed the potential of using sugars in this seaweed as feedstock for ABE production (Potts et al. 2012), but product levels were relatively low. A different study using *Ulva lactuca* from the North Sea as feedstock for ABE by other solventogenic strains demonstrated the production of ABE and 1,2-propanediol by *Clostridium beijerinckii* from rhamnose (the major sugar in *U. lactuca*) (van der Wal et al. 2013).

8. Methane Production from Anaerobic Digestion of Seaweed

The last decade has seen a renewed interest in exploiting seaweed as a feedstock in anaerobic digestion (AD). There is growing evidence to support its economic viability, and its social and environmental benefits (Costa et al. 2012, Hinks et al. 2013, Nkemka and Murto 2010, Vanegas and Bartlett 2013, Vergara-Fernández et al. 2008). Seaweeds have several characteristics which make them an ideal feedstock for AD: no lignin and low cellulose content; a high carbohydrate content that makes them easily digestible; and coastal abundance and ready availability, which allows for relatively easy collection (Jard et al. 2013, Nkemka and Murto 2010). Methane production derived from anaerobic digestion of seaweed has been demonstrated to be technically viable, and is considered competitive with other biomass energy sources in terms of cost and efficiencies (Adams et al. 2011a, Allen et al. 2016, Migliore et al. 2012).

Anaerobic digestion involves the degradation of organic matter by a diverse group of microbes in the absence of oxygen. This occurs through a series of pathways including hydrolysis, acidogenesis, acetogenesis, and methanogenesis, and ultimately produces methane and carbon dioxide (Bouallagui et al. 2009, Costa et al. 2012; Fig. 4.12). The resultant methane can be used as an alternative to natural gas, used for heat and electricity generation, or compressed for use as a transport fuel (Kelly and Dworjanyn 2008).

Various species of seaweed have been tested for methane production by AD, including *Macrocystis*, *Laminaria*, *Gracilaria*, *Sargassum*, and *Ulva* (Adams et al. 2011a, Chynoweth et al. 2001, Ghadiryanfar et al. 2016, Hinks et al. 2013, Jard et al. 2013). Allen et al. (2016) compared the biochemical methane potential (an assay used to determine the methane yield from organic matter by using an anaerobic reactor) of a range of first-, second-, and third-generation biofuels. Many seaweeds (101–225 CH_4 kg^{-1} VS; *Fucus*

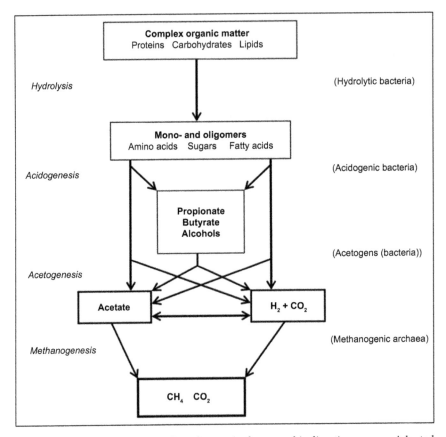

Figure 4.12. Schematic representation of stages in the anaerobic digestion process. Adapted from Demirel and Scherer 2008, Horn 2000. Stages of the process are in italics (left) while the type of microorganisms involved are in parentheses (right).

serratus and dried *Ulva lactuca,* respectively) were comparable with other biofuels (first-generation crops 306–646 CH_4 kg^{-1} VS; sugar beet tops and macerated oil seed rape, respectively; second-generation biofuels 99–804 CH_4 kg^{-1} VS; pig slurry and used cooking oil, respectively).

Some of the most important factors that affect the anaerobic digestion of seaweeds are discussed below.

1. *Pre-treatment.* Pre-treatments are commonly used to improve AD performance. These include both physical (washing, drying, and particle size reduction) and chemical treatments (Rodriguez et al. 2015). Washing is commonly used to remove particles attached to the seaweed (Adams et al. 2011a, 2015, Allen et al. 2013, Vergara-Fernández et al. 2008), but removal can have an adverse effect on the soluble

carbohydrates, thereby reducing the potential fuel content (Adams et al. 2015). Adams et al. (2015) found that higher concentrations of ethanol were produced from unwashed seaweed. However, methane concentration increased with washing, perhaps due to the impact the salt content of the washed versus unwashed seaweed had on the inoculum.

2. *Reactor Conditions (e.g., temperature)*. AD reactors are most commonly operated at either mesophilic (normally 35–47°C) or thermophilic temperatures (normally 55°C), although studies have found contradictory evidence for optimum temperatures. For example, Raposo et al. (2012) and Vanegas and Bartlett (2013) concluded that mesophilic temperatures (35°C) were optimum, while Tedesco et al. (2014) found that thermophilic temperatures (50°C) provided the most methane. Other important reactor conditions include pH and retention time (longer retention times generally produce higher methane yields; McKennedy and Sherlock 2015, Raposo et al. 2012).

3. *Inoculum and Co-digestion*. Degradation of seaweed requires a suitable inoculum, which provides the source of microorganisms. Suitable inoculums can be sourced from any environment where anaerobic methanogens are active, however different inoculum sources could lead to different degrees of performance due to their microbial and chemical compositions (Raposo et al. 2012) and marine sediment (Miura et al. 2014, Miura et al. 2015). Co-digestion with other substrates is also worth considering; it can improve the performance of difficult to degrade substrates. For example, Allen et al. (2014) found that co-digestion of dairy waste with fresh *Ulva*, a typically problematic substrate due to high concentration of sulfur and low levels of carbon to nitrogen, improved methane production relative to the substrate alone.

4. *Impact of Different Seaweed Species and Variable Composition*. As mentioned earlier, the composition of seaweeds varies by season (Adams et al. 2011b), species (Jard et al. 2013), and geographic region (Schiener et al. 2015). This can impact methane production, as the process ultimately depends on the composition of the substrate. Adams et al. (2011b) found that the highest amount of methane produced by *Laminaria digitata* (254 cm^3/g VS added) was observed in seaweed collected in July, when the carbohydrates laminarin and mannitol were at their highest; the lowest amount of methane (197 cm^3/g VS added) was produced from the seaweeds harvested in March, when the carbohydrates were at their lowest (Fig. 4.13). They suggested that blending the seaweed collected at the end with that of the beginning of the year could improve overall performance. The presence of inhibiting materials (e.g., high sulfur content, salt, or polyphenols) also impacts performance. *Sargassum*

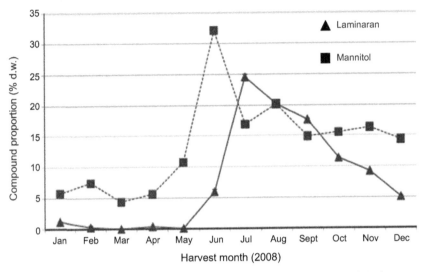

Figure 4.13. Seasonal variation in carbohydrate composition (percent dry weight) of *Laminaria digitata* (Adams et al. 2011a).

has been shown to have a high polyphenol content and associated low methane production in AD (McKennedy and Sherlock 2015).

5. *Combining Anaerobic Digestion with Other Extraction Processes.* Kerner et al. (1991) assessed the potential for AD of waste generated after the removal of other commercially viable components, such as alginate from seaweed. They found that methane could be generated from the waste products but co-digestion was required.

9. Conclusions

Macroalgae have been used since ancient times as food, fertilizer, and a source of chemicals, pharmaceuticals, and food ingredients, mainly in Asia. In recent years, macroalgae have attracted much attention for their use as feedstock for biofuel production, mainly due to the need to find renewable resources for energy and chemical production that do not compete for arable land and with food applications. To produce biofuels, macroalgae need to be cultivated at a large scale in economically viable systems. These systems are not yet available, but developments are being made towards new environmentally friendly cultivation technologies and mechanized harvest. These are expected to result in a breakthrough for economical macroalgae production (Section 2).

When combined with aquaculture in IMTA systems, the cultivation of macroalgae has important environmental and potential economic benefits. It reduces the negative environmental effects of aquaculture (i.e., eutrophication) and expands the product portfolio of sea farms. Therefore, collaboration between different sectors (including aquaculture, macroalgae cultivation, marine environmental management, and energy) would result in new, sustainable, environmentally safe, and economical systems for the production of fish, shellfish, and macroalgae (Jiang et al. 2016).

Due to their biochemical composition (rich in carbohydrates, proteins, and minerals), seaweeds are very suitable for biorefinery. Because of the large difference between species and the variations in composition during growth, it is necessary to develop specific biorefinery schemes for each type of macroalgae, with different product portfolios. Much attention has been paid to biorefinery approaches for using brown seaweeds, especially kelps, to produce fuels from carbohydrates and chemicals and feed from the residual streams. There are some reports on the biorefinery of the green species *Ulva*, in which the main products are fuels from fermentation of sugars and protein-rich feed supplements.

There is growing interest in using macroalgae as a source of high-value products such as pharmaceuticals, nutraceuticals, and other bioactive compounds (Schiener et al. 2015). This will increase the value of products that can be produced from seaweeds and increase the economic viability of the biorefinery. With the current low oil prices, it is difficult to obtain an economically sound process, even with improved substrate utilization capabilities, if fuels are the only products.

In general, the solubilization of sugars and fermentable components from macroalgae requires milder conditions than for lignocellulosic biomasses. These milder pre-treatments result in lower formation of inhibitors for fermentation and lower potential degradation of many value-added components that could be important co-products (e.g., antioxidants, vitamins, proteins). Sugar-rich hydrolysates from seaweed biomass have been successfully used for fermentation to ethanol and ABE. However, some issues regarding hydrolysate fermentability (e.g., high salt concentrations) still need to be solved before high yields can be obtained. Processes based on SSF and CBP have been studied using macroalgae as feedstock to produce ethanol; they have the potential to significantly reduce operating costs in a commercial fermentation process.

Macroalgae exhibit good potential for AD: both wild sources (beach coast) and farmed feedstock. They are easily degradable and combine the absence of lignin with a high carbohydrate content, making them an ideal substrate for AD (Jard et al. 2013). Methane production from seaweed is technically feasible, and is considered competitive with that from other biomass feedstocks (Allen et al. 2016). Further research and optimisation is, however, required to overcome composition and inhibition issues and to improve performance and reliability.

Technological advances in the production of biofuels from macroalgae will profit from fundamental and applied research, and from knowledge built on the utilization of first- and second-generation biomass. Topics like pre-treatment with a higher efficiency, prevention of polysaccharide degradation, and co-utilization of hexoses and pentoses are typical bottlenecks being addressed in lignocellulosic biofuel production. Many groups are studying issues specific to seaweed biomass, such as large-scale sustainable cultivation. The environmental and economic benefits that are potentially associated with IMTA systems could be especially helpful to obtaining policy support and commercializing the processes.

Acknowledgements

This work is supported by the Macrofuels European project, which received funding from the European Union's Horizon 2020 research and innovation programme under grant agreement No. 654010.

Keywords: Macroalgae; seaweeds; IMTA; biorefinery; anaerobic digestion; ABE fermentation; bioethanol; biobutanol; thermochemical conversion; HTL

References

2010. F.O. Lichts world sugar and sweetener yearbook. F.O. Licht, Ratzeburg.

Abreu, M.H., R. Pereira, C. Yarish, A.H. Buschmann and I. Sousa-Pinto. 2011. IMTA with *Gracilaria vermiculophylla*: Productivity and nutrient removal performance of the seaweed in a land-based pilot scale system. Aquaculture 312: 77–87.

Adams, J.M., J.A. Gallagher and I.S. Donnison. 2009. Fermentation study on *Saccharina latissima* for bioethanol production considering variable pre-treatments. J. Appl. Phycol. 21: 569–574.

Adams, J.M.M., A.B. Ross, K. Anastasakis, E.M. Hodgson, J.A. Gallagher, J.M. Jones et al. 2011a. Seasonal variation in the chemical composition of the bioenergy feedstock *Laminaria digitata* for thermochemical conversion. Bioresour. Technol. 102: 226–234.

Adams, J.M.M., T.A. Toop, I.S. Donnison and J.A. Gallagher. 2011b. Seasonal variation in *Laminaria digitata* and its impact on biochemical conversion routes to biofuels. Bioresour. Technol. 102: 9976–9984.

Adams, J.M.M., A. Schmidt and J.A. Gallagher. 2015. The impact of sample preparation of the macroalgae *Laminaria digitata* on the production of the biofuels bioethanol and biomethane. J. Appl. Phycol. 27: 985–991.

Allen, E., J. Browne, S. Hynes and J.D. Murphy. 2013. The potential of algae blooms to produce renewable gaseous fuel. Waste Management 33: 2425–2433.

Allen, E., D.M. Wall, C. Herrmann and J.D. Murphy. 2014. Investigation of the optimal percentage of green seaweed that may be co-digested with dairy slurry to produce gaseous biofuel. Bioresour. Technol. 170: 436–444.

Allen, E., D.M. Wall, C. Herrmann and J.D. Murphy. 2016. A detailed assessment of resource of biomethane from first, second and third generation substrates. Renew Energy 87: 656–665.

Almarsdottir, A.R., M.A. Sigurbjornsdottir and J. Orlygsson. 2012. Effect of various factors on ethanol yields from lignocellulosic biomass by *Thermoanaerobacterium* AK17. Biotechnol. Bioeng. 109: 686–694.

Anastasakis, K. and A.B. Ross. 2011. Hydrothermal liquefaction of the brown macro-alga *Laminaria saccharina*: Effect of reaction conditions on product distribution and composition. Bioresour. Technol. 102: 4876–4883.

Anastasakis, K., A.B. Ross and J.M. Jones. 2011. Pyrolysis behaviour of the main carbohydrates of brown macro-algae. Fuel 90: 598–607.

Anastasakis, K. and A.B. Ross. 2015. Hydrothermal liquefaction of four brown macro-algae commonly found on the UK coasts: An energetic analysis of the process and comparison with bio-chemical conversion methods. Fuel 139: 546–553.

Bach, Q.V., M.V. Sillero, K.Q. Tran and J. Skjermo. 2014. Fast hydrothermal liquefaction of a Norwegian macro-alga: Screening tests. Algal Res. 6, Part B: 271–276.

Baghel, R.S., P. Kumari, C.R.K. Reddy and B. Jha. 2014. Growth, pigments, and biochemical composition of marine red alga *Gracilaria crassa*. J. Appl. Phycol. 26: 2143–2150.

Barrington, K., T. Chopin and S. Robinson. 2009. Integrated multi-trophic aquaculture (IMTA) in marine temperate waters. *In:* D. Soto (ed.). Integrated Marine Aquaculture: A Global Review. FAO Fisheries and Aquaculture Technical Paper. FAO. Rome.

Bikker, P., M.M. van Krimpen, P. van Wikselaar, B. Houweling-Tan, N. Scaccia, J.W. van Hal et al. 2016. Biorefinery of the green seaweed *Ulva lactuca* to produce chemicals, biofuels and animal feed. J. Appl. Phycol. 28(6): 3511–3525.

Bouallagui, H., H. Lahdheb, E. Ben Romdan, B. Rachdi and M. Hamdi. 2009. Improvement of fruit and vegetable waste anaerobic digestion performance and stability with co-substrates addition. J. Environ. Management 90: 1844–1849.

Brennan, L. and P. Owende. 2010. Biofuels from microalgae—A review of technologies for production, processing, and extractions of biofuels and co-products. Renew. Sust. Energy Rev. 14: 557–577.

Bruton, T., H. Lyons, Y. Lerat, M. Stanley and M. Bo Rasmussen. 2009. A review of the potential of marine algae as a source of biofuel in Ireland. Sustainable Energy Ireland.

Chang, T.H. and S. Yao. 2011. Thermophilic, lignocellulolytic bacteria for ethanol production: current state and perspectives. Appl. Microbiol. Biotechnol. 92: 13–27.

Chen, H., D. Zhou, G. Luo, S. Zhang and J. Chen. 2015. Macroalgae for biofuels production: Progress and perspectives. Renew. Sust. Energy Rev. 47: 427–437.

Cho, Y., H. Kim and S.K. Kim. 2013. Bioethanol production from brown seaweed, *Undaria pinnatifida*, using NaCl acclimated yeast. Bioprocess Biosyst. Eng. 36: 713–719.

Chung, D., M. Cha, A.M. Guss and J. Westpheling. 2014. Direct conversion of plant biomass to ethanol by engineered *Caldicellulosiruptor bescii*. Proceedings Nat. Acad. Sci. (PNAS) 111: 8931–8936.

Chynoweth, D.P., J.M. Owens and R. Legrand. 2001. Renewable methane from anaerobic digestion of biomass. Renew. Energy 22: 1–8.

Costa, J.C., P.R. Gonçalves, A. Nobre and M.M. Alves. 2012. Biomethanation potential of macroalgae *Ulva* spp. and *Gracilaria* spp. and in co-digestion with waste activated sludge. Bioresour. Technol. 114: 320–326.

Cunha, L. and A. Grenha. 2016. Sulfated seaweed polysaccharides as multifunctional materials in drug delivery applications. Marine Drugs 14: 42.

Demirel, B. and P. Scherer. 2008. The roles of acetotrophic and hydrogenotrophic methanogens during anaerobic conversion of biomass to methane: A review. Rev. Environ. Sci. Biotechnol. 7: 173–190.

Elliott, D.C., T.R. Hart, G.G. Neuenschwander, L.J. Rotness, G. Roesijadi, A.H. Zacher et al. 2013. Hydrothermal processing of macroalgal feedstocks in continuous-flow reactors. ACS Sust. Chem. Eng. 2: 207–215.

Elliott, D.C. 2016. Review of recent reports on process technology for thermochemical conversion of whole algae to liquid fuels. Algal Res. 13: 255–263.

Enquist-Newman, M., A.M.E. Faust, D.D. Bravo, C.N.S. Santos, R.M. Raisner, A. Hanel et al. 2014. Efficient ethanol production from brown macroalgae sugars by a synthetic yeast platform. Nature 505: 239–243.

FAO. 2012. Fishery and Aquaculture Statistics 2010.

Fei, X.G. 2004. Solving the coastal eutrophication problem by large scale seaweed cultivation. Hydrobiologia 512.

Fleurence, J. 1999. Seaweed proteins: Biochemical, nutritional aspects and potential uses. Trends Food Sci. Technol. 10: 25–28.

Florentinus, A., C. Hamelinck, S. de Lint and S. van Iersel. 2014. Worldwide potential of aquatic biomass. Ecofys Bio Energy Group (http://www.ecofys.com/files/files/ecofys-2008-worldwide-potential-of-aquatic-biomass-revision-2014.pdf), Utrecht.

Galland-Irmouli, A.V.r., J.l. Fleurence, R. Lamghari, M. Luçon, C. Rouxel, O. Barbaroux et al. 1999. Nutritional value of proteins from edible seaweed *Palmaria palmata* (dulse). J. Nutritional Biochem. 10: 353–359.

Gaserod, O. 2011. Oral immunostimulation of mammals, birds and reptiles from (1–4) linked beta D mannuronic acid.

Ghadiryanfar, M., K.A. Rosentrater, A. Keyhani and M. Omid. 2016. A review of macroalgae production, with potential applications in biofuels and bioenergy. Renew. Sust. Energy Rev. 54: 473–481.

Graiff, A., W. Ruth, U. Kragl and U. Karsten. 2016. Chemical characterization and quantification of the brown algal storage compound laminarin—A new methodological approach. J. Appl. Phycol. 28: 533–543.

Green, E.M. 2011. Fermentative production of butanol: the industrial perspective. Curr. Opinion Biotechnol. 22: 337–343.

Groenendijk, F.e., P. Bikker, R. Blaauw, W. Brandenburg, S.v.d. Burg, L.v. Duren et al. 2016. North-Sea-Weed-Chain. Sustainable seaweed from the North Sea; an exploration of the value chain (http://www.wageningenur.nl/upload_mm/8/3/a/5d951efe-9446-45c1-a52a-b7c46679d0f7_C055.16%20NorthSeaWeedChain%20-%20Floris%20Groenendijk.pdf). WUR-Imares, Wageningen.

Hargreaves, P.I., C.A. Barcelos, A.C.A. da Costa and N. Pereira, Jr. 2013. Production of ethanol 3G from *Kappaphycus alvarezii:* Evaluation of different process strategies. Bioresour. Technol. 134: 257–263.

Harmsen, P.F.H., M.M. Hackmann and H.L. Bos. 2014. Green building blocks for bio-based plastics. Biofuels Bioprod. Biorefin. 8: 306–324.

He, Y., X. Liang, C. Jazrawi, A. Montoya, A. Yuen, A.J. Cole et al. 2016. Continuous hydrothermal liquefaction of macroalgae in the presence of organic co-solvents. Algal Res. 17: 185–195.

Hinks, J., S. Edwards, P.J. Sallis and G.S. Caldwell. 2013. The steady state anaerobic digestion of *Laminaria hyperborea*—Effect of hydraulic residence on biogas production and bacterial community composition. Bioresour. Technol. 143: 221–230.

Holdt, S.L. and S. Kraan. 2011. Bioactive compounds in seaweed: functional food applications and legislation. J. Appl. Phycol. 23: 543–597.

Horn, S.J. 2000. Bioenergy from brown seaweeds. Sci. Technol. November: Department of Biotechnology, Norwegian University.

Horn, S.J., I.M. Aasen and K. Ostgaard. 2000a. Ethanol production from seaweed extract. J. Ind. Microbiol. Biot. 25: 249–254.

Horn, S.J., I.M. Aasen and K. Ostgaard. 2000b. Production of ethanol from mannitol by Zymobacter palmae. J. Ind. Microbiol. Biot. 24: 51–57.

Hou, X. 2012. Anaerobic xylose fermentation by *Spathaspora passalidarum*. Appl. Microbiol. Biotechnol. 94: 205–214.

Hou, X., J.H. Hansen and A.-B. Bjerre. 2015. Integrated bioethanol and protein production from brown seaweed *Laminaria digitata*. Bioresour. Technol. 197: 310–317.

Hreggvidsson, G.O., W.J.O. Jonsson, B. Bjornsdottir, O.H. Fridjonsson, J. Altenbuchner, H. Watzlawick et al. 2015. Thermostable alginate degrading enzymes and their methods of use. WO 2015104723 A1.

Huesemann, M.H., L.-J. Kuo, L. Urquhart, G.A. Gill and G. Roesijadi. 2012. Acetone-butanol fermentation of marine macroalgae. Bioresour. Technol. 108: 305–309.

Jang, S.S., Y. Shirai, M. Uchida and M. Wakisaka. 2012. Production of mono sugar from acid hydrolysis of seaweed. African J. Biotech. 11: 1953–1963.

Jard, G., H. Marfaing, H. Carrère, J.P. Delgenes, J.P. Steyer and C. Dumas. 2013. French Brittany macroalgae screening: Composition and methane potential for potential alternative sources of energy and products. Bioresour. Technol.

Jeon, W., C. Ban, G. Park, T.K. Yu, J.Y. Suh, H.C. Woo et al. 2015. Catalytic hydrothermal conversion of macroalgae-derived alginate: Effect of pH on production of furfural and valuable organic acids under subcritical water conditions. J. Mol. Catalysis A: Chemical 399: 106–113.

Jeon, W., C. Ban, G. Park, H.C. Woo and D.H. Kim. 2016. Hydrothermal conversion of alginic acid to furfural catalyzed by Cu(II) ion. Catal. Today 265: 154–162.

Ji, S.Q., B. Wang, M. Lu and F.L. Li. 2016. Direct bioconversion of brown algae into ethanol by thermophilic bacterium *Defluviitalea phaphyphila*. Biotechnol. Biofuels 9.

Jiang, R., K.N. Ingle and A. Golberg. 2016. Macroalgae (seaweed) for liquid transportation biofuel production: what is next? Algal Res. 14: 48–57.

Jung, K.A., S.-R. Lim, Y. Kim and J.M. Park. 2013. Potentials of macroalgae as feedstocks for biorefinery. Bioresour. Technol. 135: 182–190.

Kelly, M.S. and S. Dworjanyn. 2008. The potential of marine biomass for anaerobic biogas production: A feasibility study with recommendations for further reasearch. The Crown Estate Scottish Association for Marine Science.

Kerner, K.N., J.F. Hanssen and T.A. Pedersen. 1991. Anaerobic digestion of waste sludges from the alginate extraction process. Bioresour. Technol. 37: 17–24.

Kim, H.S., C.-G. Lee and E.Y. Lee. 2011. Alginate lyase: Structure, property, and application. Biotechnol. Bioprocess Eng. 16: 843–851.

Klinke, H.B., L. Olsson, A.B. Thomsen and B.K. Ahring. 2003. Potential inhibitors from wet oxidation of wheat straw and their effect on ethanol production of *Saccharomyces cerevisiae*: Wet oxidation and fermentation by yeast. Biotechnol. Bioeng. 81: 738–747.

Korzen, L., I.N. Pulidindi, A. Israel, A. Abelson and A. Gedanken. 2015. Single step production of bioethanol from the seaweed *Ulva rigida* using sonication. RSC Adv. 5: 16223–16229.

Koskinen, P.E.P., S.R. Beck, J. Orlygsson and J.A. Puhakka. 2008. Ethanol and hydrogen production by two thermophilic, anaerobic bacteria isolated from icelandic geothermal areas. Biotechnol. Bioeng. 101: 679–690.

Kraan, S. 2012. Algal polysaccharides, novel applications and outlook. *In*: Chuan-Fa Chang (ed.). Carbohydrates—Comprehensive Studies on Glycobiology and Glycotechnology. InTech, Doi: 10.5772/51572.

Kraan, S. 2013. Mass-cultivation of carbohydrate rich macroalgae, a possible solution for sustainable biofuel production. 18: 27–46.

Kumar, S., R. Gupta, G. Kumar, D. Sahoo and R.C. Kuhad. 2013. Bioethanol production from *Gracilaria verrucosa*, a red alga, in a biorefinery approach. Bioresour. Technol. 135: 150–156.

Lahaye, M. and J. Vigouroux. 1992. Liquefaction of dulse (*Palmaria palmata* (L.) Kuntze) by a commercial enzyme preparation and a purified endo,+Ý-1,4-D-xylanase. J. Appl. Phycol. 4: 329–337.

Lahaye, M., C. Michel and J.L. Barry. 1993. Chemical, physicochemical and *in-vitro* fermentation characteristics of dietary fibers from *Palmaria palmata* (L.) Kuntze. Food Chem. 47: 29–36.

Lahaye, M., C. Rondeau-Mouro, E. Deniaud and A. Buléon. 2003. Solid-state 13C NMR spectroscopy studies of xylans in the cell wall of *Palmaria palmata* (L. Kuntze, Rhodophyta). Carbohydrate Res. 338: 1559–1569.

Lahaye, M. and A. Robic. 2007. Structure and functional properties of Ulvan, a polysaccharide from green seaweeds. Biomacromolecules 8: 1765–1774.

Lange, J.P., E. van der Heide, J. van Buijtenen and R. Price. 2012. Furfural: Promising platform for lignocellulosic biofuels. Chem. Sus. Chem. 5: 150–166.

Lee, J.Y., P. Li, J. Lee, H.J. Ryu and K.K. Oh. 2013. Ethanol production from *Saccharina japonica* using an optimized extremely low acid pretreatment followed by simultaneous saccharification and fermentation. Bioresour. Technol. 127: 119–125.

Lewis, J. 2011. Product options for the processing of marine macro algae-summary report. The Crown Estate, The Centre for process innovation.

López Barreiro, D., M. Beck, U. Hornung, F. Ronsse, A. Kruse and W. Prins. 2015. Suitability of hydrothermal liquefaction as a conversion route to produce biofuels from macroalgae. Algal Res. 11: 234–241.

López-Contreras, A.M., H. Smidt, J. van der Oost, P.A.M. Claassen, H. Mooibroek and W.M. de Vos. 2001. *Clostridium beijerinckii* cells expressing *Neocallimastix patriciarum* glycoside hydrolases show enhanced lichenan utilization and solvent production. Appl. Envron. Microbiol. 67: 5127–5133.

López Contreras, A.M., W. Kuit, M.A.J. Siemerink, S.W.M. Kengen, J. Springer and P.A.M. Claassen. 2010. Production of longer-chain alcohols from biomass—butanol, isopropanol and 2,3-butanediol. *In*: Bioalcohol Production: Biochemical Conversion of Lignocellulosic Biomass. Woodhead Publishing Ltd. (ISBN 9781845695101) (Doi: urn:nbn:nl:ui:32-393273).

López-Contreras, A.M., P.F.H. Harmsen, R. Blaauw, B. Howeling-Tan, H. van der Wal, W. Huijgen et al. 2014. Biorefinery of the brown seaweed *Saccharina latissima* for fuels and chemicals. Mie Bioforum on lignocellulose degradation and biorefinery, Nemunosato Resort, Japan.

Lovell, R.T. 2003. 13—Diet and Fish Husbandry A2—Halver, John E. 703–754. *In*: R.W. Hardy (ed.). Fish Nutrition (Third Edition). Academic Press. San Diego.

Lynd, L.R. 1996. Overview and evaluation of fuel ethanol from cellulosic biomass: Technology, economics, the environment and policy. Ann. Rev. Energy Environment 21: 403–465.

Mariscal, R., P. Maireles-Torres, M. Ojeda, I. Sadaba and M. Lopez Granados. 2016. Furfural: a renewable and versatile platform molecule for the synthesis of chemicals and fuels. Energy Environ. Sci. 9: 1144–1189.

McKennedy, J. and O. Sherlock. 2015. Anaerobic digestion of marine macroalgae: A review. Renew. Sust. Energy Rev. 52: 1781–1790.

Migliore, G., C. Alisi, a.R. Sprocati, E. Massi, R. Ciccoli, M. Lenzi et al. 2012. Anaerobic digestion of macroalgal biomass and sediments sourced from the Orbetello lagoon, Italy. Biomass 42: 69–77.

Milledge, J.J., B. Smith, P.W. Dyer and P. Harvey. 2014. Macroalgae-derived biofuel: A review of methods of energy extraction from seaweed biomass. Energies 7: 7194–7222.

Miura, T., A. Kita, Y. Okamura, T. Aki, Y. Matsumura, T. Tajima et al. 2014. Evaluation of marine sediments as microbial sources for methane production from brown algae under high salinity. Bioresour. Technol. 169: 362–366.

Miura, T., A. Kita, Y. Okamura, T. Aki, Y. Matsumura, T. Tajima et al. 2015. Improved methane production from brown algae under high salinity by fed-batch acclimation. Bioresour. Technol. 187: 275–281.

Morgan, K., J. Wright and F.J. Simpson. 1980. Review of chemical constituents of the red alga *Palmaria palmata* (dulse). Econ. Bot. 34: 27–50.

Mutripah, S., M.D.N. Meinita, J.-Y. Kang, G.-T. Jeong, A. Susanto, R.E. Prabowo et al. 2014. Bioethanol production from the hydrolysate of *Palmaria palmata* using sulfuric acid and fermentation with brewer's yeast. J. Appl. Phycol. 26: 687–693.

Neori, A., T. Chopin, M. Troell, A.H. Buschmann, G.P. Kraemer, C. Halling et al. 2004. Integrated aquaculture: Rationale, evolution and state of the art emphasizing seaweed biofiltration in modern mariculture. Aquaculture 231: 361–391.

Neveux, N., A.K.L. Yuen, C. Jazrawi, M. Magnusson, B.S. Haynes, A.F. Masters et al. 2014. Biocrude yield and productivity from the hydrothermal liquefaction of marine and freshwater green macroalgae. Bioresour. Technol. 155: 334–341.

Nielsen, M.M., D. Manns, M. D'Este, D. Krause-Jensen, M.B. Rasmussen, M.M. Larsen et al. 2016. Variation in biochemical composition of *Saccharina latissima* and *Laminaria digitata* along an estuarine salinity gradient in inner Danish waters. Algal Res. 13: 235–245.

Nkemka, V.N. and M. Murto. 2010. Evaluation of biogas production from seaweed in batch tests and in UASB reactors combined with the removal of heavy metals. J. Environ. Management 91: 1573–1579.

Nunes, A.J.P., M.V.C. Sá, C.L. Browdy and M. Vazquez-Anon. 2014. Practical supplementation of shrimp and fish feeds with crystalline amino acids. Aquaculture 431: 20–27.

Olofsson, K., M. Bertilsson and G. Lidén. 2008. A short review on SSF—an interesting process option for ethanol production from lignocellulosic feedstocks. Biotechnol. Biofuels 1: 7.

Palmqvist, E. and B. Hahn-Hägerdal. 2000. Fermentation of lignocellulosic hydrolysates. II: Inhibitors and mechanisms of inhibition. Bioresour. Technol. 74: 25–33.

Pangestuti, R. and S.-K. Kim. 2011. Neuroprotective effects of marine algae. Marine Drugs 9: 803.

Park, G., W. Jeon, C. Ban, H.C. Woo and D.H. Kim. 2016. Direct catalytic conversion of brown seaweed-derived alginic acid to furfural using 12-tungstophosphoric acid catalyst in tetrahydrofuran/water co-solvent. Energy Convers. Management 118: 135–141.

Philippidis, G.P., T.K. Smith and C.E. Wyman. 1993. Study of the enzymatic hydrolysis of cellulose for production of fuel ethanol by the simultaneous saccharification and fermentation process. Biotechnol. Bioeng. 41: 846–853.

Potts, T., J. Du, M. Paul, P. May, R. Beitle and J. Hestekin. 2012. The production of butanol from Jamaica bay macro algae. Environ. Prog. Sustainable Energy 31: 29–36.

Raposo, F., M.A. De la Rubia, V. Fernández-Cegrí and R. Borja. 2012. Anaerobic digestion of solid organic substrates in batch mode: An overview relating to methane yields and experimental procedures. Renew. Sust. Energy Rev. 16: 861–877.

Reece, J.B. and N.A. Campbell. 2011. Campbell biology. Benjamin Cummings/Pearson, Boston.

Reith, J.H., E.P. Deurwaarder, A.P.W.M. Curvers, P. Kamersmans and W. Brandenburg. 2005. Bio-offshore: Grootschalige teelt van zeewieren in combinatie met off-shore windparken in de Noordzee.

Rodriguez, C., A. Alaswad, J. Mooney, T. Prescott and A.G. Olabi. 2015. Pre-treatment techniques used for anaerobic digestion of algae. Fuel Processing Technology 138: 765–779.

Roesijadi, G., S.B. Jones, L.J. Snowden-Swan and Y. Zhu. 2010. Macroalgae as a biomass feedstock: a preliminary analysis. Pacific Northwest National Laboratories—US department of Energy (PNNL-DOE).

Ross, A.B., J.M. Jones, M.L. Kubacki and T. Bridgeman. 2008. Classification of macroalgae as fuel and its thermochemical behaviour. Bioresour. Technol. 99: 6494–6504.

Ross, A.B., K. Anastasakis, M. Kubacki and J.M. Jones. 2009. Investigation of the pyrolysis behaviour of brown algae before and after pre-treatment using Py-GC/MS and TGA. J. Anal. Appl. Pyrol.: 3–10.

Rudolf, A., M. Alkasrawi, G. Zacchi and G. Lidén. 2005. A comparison between batch and fed-batch simultaneous saccharification and fermentation of steam pretreated spruce. Enzyme Microbial Technol. 37: 195–204.

Sassner, P., M. Galbe and G. Zacchi. 2008. Techno-economic evaluation of bioethanol production from three different lignocellulosic materials. Biomass 32: 422–430.

Schiener, P., K.D. Black, M.S. Stanley and D.H. Green. 2015. The seasonal variation in the chemical composition of the kelp species *Laminaria digitata, Laminaria hyperborea, Saccharina latissima* and *Alaria esculenta*. J. Appl. Phycol. 27: 363–373.

Schultz-Jensen, N., A. Thygesen, F. Leipold, S.T. Thomsen, C. Roslander, H. Lilholt et al. 2013. Pretreatment of the macroalgae *Chaetomorpha linum* for the production of bioethanol— Comparison of five pretreatment technologies. Bioresour. Technol. 140: 36–42.

Scully, S. and J. Orlygsson. 2015. Recent advances in second generation ethanol production by thermophilic bacteria. Energies 8: 1.

Shahsavarani, H., D. Hasegawa, D. Yokota, M. Sugiyama, Y. Kaneko, C. Boonchird et al. 2013. Enhanced bio-ethanol production from cellulosic materials by semi-simultaneous saccharification and fermentation using high temperature resistant *Saccharomyces cerevisiae* TJ14. J. Biosci. Bioeng. 115: 20–23.

Suganya, T., M. Varman, H.H. Masjuki and S. Renganathan. 2016. Macroalgae and microalgae as a potential source for commercial applications along with biofuels production: A biorefinery approach. Renew. Sust. Energy Rev. 55: 909–941.

Suutari, M., E. Leskinen, K. Fagerstedt, J. Kuparinen, P. Kuuppo and J. Blomster. 2015. Macroalgae in biofuel production. Phycol. Res. 63: 1–18.

Sveinsdottir, M., S.R.B. Baldursson and J. Orlygsson. 2009. Ethanol production from monosugars and lignocellulosic biomass by thermophilic bacteria isolated from Icelandic hot springs. Iceland Agr. Sci. 22: 45–58.

Takagi, M., S. Abe, S. Suzuki, G.H. Evert and N. Yata. 1977. A method of production of alcohol directly from yeast. Proceedings of Bioconversion Symposium.

Takagi, T., M. Uchida, R. Matsushima, H. Kodama, T. Takeda, M. Ishida et al. 2015. Comparison of ethanol productivity among yeast strains using three different seaweeds. Fisheries Sci. 81: 763–770.

Tedesco, S., D. Mac Lochlainn and A.G. Olabi. 2014. Particle size reduction optimization of *Laminaria* spp. biomass for enhanced methane production. Energy 76: 857–862.

Thygesen, A., A.B. Thomsen, A.S. Schmidt, H. Jørgensen, B.K. Ahring and L. Olsson. 2003. Production of cellulose and hemicellulose-degrading enzymes by filamentous fungi cultivated on wet-oxidised wheat straw. Enzyme Microbial Technol. 32: 606–615.

Tian, L., B. Papanek, D.G. Olson, T. Rydzak, E.K. Holwerda, T. Zheng et al. 2016. Simultaneous achievement of high ethanol yield and titer in *Clostridium thermocellum* Biotech. Biofuels 9.

Troell, M., C. Halling, A. Neori, T. Chopin, A.H. Buschmann, N. Kautsky et al. 2003. Integrated mariculture: asking the right questions. Aquaculture 226: 69–90.

Turner, P., G. Mamo and E.N. Karlsson. 2007. Potential and utilization of thermophiles and thermostable enzymes in biorefining. Microbial Cell Factories 6: 1–23.

Valente, L.M.P., A. Gouveia, P. Rema, J. Matos, E.F. Gomes and I.S. Pinto. 2006. Evaluation of three seaweeds *Gracilaria bursa-pastoris*, *Ulva rigida* and *Gracilaria cornea* as dietary ingredients in European sea bass (*Dicentrarchus labrax*) juveniles. Aquaculture 252: 85–91.

van den Burg, S., M. Stuiver, F. Veenstra and P. Bikker. 2013. A triple P review of the feasibility of sustainable offshore seaweed production in the North Sea. Wageningen UR, Wageningen, the Netherlands.

van der Wal, H., B. Sperber, B. Houweling-Tan, R. Bakker, W. Brandenburg and A. Lopez-Contreras. 2013. Production of acetone, butanol, and ethanol from biomass of the green seaweed *Ulva lactuca*. Bioresour. Technol. 128: 431–437.

van Hal, J.W., W.J.J. Huijgen and A.M. López-Contreras. 2014. Opportunities and challenges for seaweed in the biobased economy. Trends Biotechnol. 32: 231–233.

van Maris, A.J.A., D.A. Abbott, E. Bellissimi, J. van den Brink, M. Kuyper, M.A.H. Luttik et al. 2006. Alcoholic fermentation of carbon sources in biomass hydrolysates by *Saccharomyces cerevisiae*: current status. Antonie van Leeuwenhoek 90: 391–418.

Van Putten, R.J., J.C.v.d. Waal, E. Jong de, C.B. Rasrendra, H.J. Heeres and J.G. De Vries. 2013. Hydroxymethylfurfural, a versatile platform chemical made from renewable resources. Chem. Rev.

Vanegas, C.H. and J. Bartlett. 2013. Green energy from marine algae: Biogas production and composition from the anaerobic digestion of Irish seaweed species. Environ. Technol. 34: 2277–2283.

Vergara-Fernández, A., G. Vargas, N. Alarcón and A. Velasco. 2008. Evaluation of marine algae as a source of biogas in a two-stage anaerobic reactor system. Biomass 32: 338–344.

Villanueva, R.D., A.M.M. Sousa, M.P. Gonçalves, M. Nilsson and L. Hilliou. 2010. Production and properties of agar from the invasive marine alga, *Gracilaria vermiculophylla* (Gracilariales, Rhodophyta). J. Appl. Phycol. 22: 211–220.

Wang, X., X. Liu and G. Wang. 2011. Two-stage hydrolysis of invasive algal feedstock for ethanol fermentation. J. Integr. Plant Biol. 53: 246–252.

Wargacki, A.J., E. Leonard, M.N. Win, D.D. Regitsky, C.N.S. Santos, P.B. Kim et al. 2012. An engineered microbial platform for direct biofuel production from brown macroalgae. Science 335: 308–313.

Wei, N., J. Quarterman and Y.-S. Jin. 2013. Marine macroalgae: an untapped resource for producing fuels and chemicals. Trends Biotechnol. 31: 70–77.

Wright, J.D., C.E. Wyman and K. Grohmann. 1988. Simultaneous saccharification and fermentation of lignocellulose—Process evaluation. Appl. Biochem. Biotechnol. 18: 75–90.

Yanagisawa, M., K. Nakamura, O. Ariga and K. Nakasaki. 2011. Production of high concentrations of bioethanol from seaweeds that contain easily hydrolyzable polysaccharides. Process Biochem. 46: 2111–2116.

Yang, X., M. Xu and S.T. Yang. 2015. Metabolic and process engineering of *Clostridium cellulovorans* for biofuel production from cellulose. Metabolic Eng. 32: 39–48.

Zeitsch, K.J. 2010. The chemistry and technology of furfural and its many by-products. Elsevier.

CHAPTER 5

Thermochemical Treatment

Filomena Pinto,[1,]* Paula Costa and *Miguel Miranda*

1. Introduction

The use of algal biomass for biofuels production has many advantages, namely: algal biomass can be produced all over the year and have a rapid growth capability, it grows in aqueous media, but it requires less water than terrestrial crops, it can be grown in brackish water on non-arable land, algae cultivation does not need herbicides or pesticides application and the nutrients for algae growth (mainly nitrogen and phosphorus) can be obtained from wastewater, thus as the algae grow, water effluents from agro-industrial sectors are treated. Algae growth does not compete with food production and it improves air quality due to CO_2 biofixation, as 1 kg of dry algal biomass use around 1.83 kg of CO_2 (Brennan and Owende 2010). Many species of microalgae have oil content between 20 and 50% dry weight, and by changing the growth conditions the oil yield may increase significantly.

The thermochemical processes developed for biomass energetic valorisation may also be used for algal biomass, keeping in mind the specificities of this type of biomass.

Thermochemical processes are usually divided into dry or conventional, and wet or new hydrothermal processes that operate under sub or supercritical conditions. In Fig. 5.1, a schematic diagram about the main products obtained by different thermochemical processes is presented.

Conventional or dry thermochemical processes require low water contents, usually below 20% and thus, they are not suitable for feedstocks

[1] LNEG, Estrada do Paço do Lumiar, 22, 1649-038 Lisboa, PORTUGAL.
* Corresponding author: filomena.pinto@lneg.pt

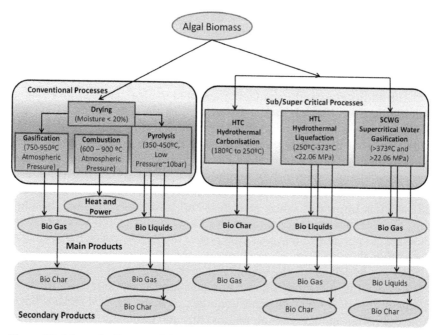

Figure 5.1. Schematic diagram of the main products obtained by different thermochemical processes applied to algal biomass.

with higher water contents as it happens with algal biomass. On the other hand, the new hydrothermal or wet processes are suitable for feedstocks with high water contents, as water is usually added to the process to act as a solvent and/or a catalyst. There are mainly three processes, hydrothermal carbonization and hydrothermal liquefaction that both occur at water sub-critical condition, while supercritical water gasification requires water super-critical conditions. Both conventional and new hydrothermal processes will be analysed next with some detail.

2. Thermochemical Conventional Processes

2.1 Introduction

Conventional or dry thermochemical processes are suitable for feedstocks with lower water content, generally below 20%. Conventional thermochemical processes include direct combustion, gasification, liquefaction, and pyrolysis. As shown in Fig. 5.1, the main objective of direct combustion is the production of heat and power, while for gasification is the production of gases. The main aim of liquefaction and pyrolysis processes is the production of liquid bio oils, though some gases and biochar are also produced. Combustion is the most mature technology. The second most mature technology is gasification, followed by pyrolysis. Conventional thermochemical processes are not suitable for algal biomass, because all of them require low water contents and thus, high energy demanding processes for algae concentration and drying processes are required.

The additional energy demand associated to algal biomass processing before direct combustion makes this process inadequate for algal biomass valorisation. As mentioned by Brennan and Owende 2010, there is little indication that direct combustion of algal biomass is technically viable. Nevertheless, co-firing of coal and algae mixtures or of other biomass feedstocks blended with algae could lead to lower emissions of greenhouse gases and favourable life cycle assessment. The overall combustion plant efficiency is improved when both heat and power is generated, besides this, higher efficiencies are obtained for larger plants (> 100 MW) or when biomass is co-combusted in coal fired power plants. However, the available data about algae co-firing is very limited and further research of this subject is required to determine the viability of such processes.

Besides the energy demands problems associated to the use of algal biomass in conventional thermochemical processes, several studies have been published about algae valorisation through gasification and pyrolysis processes, mainly pyrolysis. Thus, some information about these processes will be presented next.

2.2 Gasification

Gasification converts biomass into a gas, also referred as syngas, whose major compounds are: carbon monoxide (CO), carbon dioxide (CO_2), hydrogen (H_2), methane (CH_4), and other gaseous hydrocarbons from C_2 to C_4, usually referred as (C_nH_m). Gasification gas or syngas may have different applications, the most common one is as biofuel for energy production. Boilers, motor, or turbines may be used for this purpose. Gasification gas may be also used in chemical synthesis to produce liquid or gaseous fuels. Before being suitable for chemical synthesis, syngas usually needs cleaning and upgrading to achieve the very low required contents of tar, alkali metal and sulphur, nitrogen and chlorine compounds. In addition, the H_2/CO ratio in the gasification gas must be between 1.5 and 3.0 depending on the chemical synthesis: Fischer-Tropsch, synthetic natural gas (SNG), alcohol syntheses (methanol, ethanol and propanol), or synthesis of DME (dimethyl ether). Due to the costs associated to the production of diesel like biofuels by Fischer-Tropsch synthesis, the later investigation work have been focusing on research and development of technologies for the production of alcohol mixtures and dimethyl ether (DME) by chemical syntheses. In Fig. 5.2, are presented the main products that may be obtained from biomass gasification.

Gasification occurs at temperatures from 750° to more than 1400°C, depending on gasification technology used, for heterogeneous and poor quality biomass, fluidised bed processes are the most adequate, because high mass and energy transfer rates are easily achieved, and thus higher conversion and gas yields are obtained. Fluidised bed gasification requires temperature in the range from 750° to 900°C, particle sizes from 1 to 5 mm, and reaction times between 5 and 50 seconds. Besides fluidised bed reactors, fixed beds and entrained flow reactors are other options. The latter have the great advantage of decreasing considerable the release of tar, but they require very low particle sizes, usually below 500 µm, gasification temperatures between 900 and 1.400°C, and residence times below 10 seconds.

During biomass heating, biomass drying with the release of water vapour from the surface and from the inner pores of the solid biomass occurs first. Next, at higher temperature there are the devolatilisation or pyrolysis

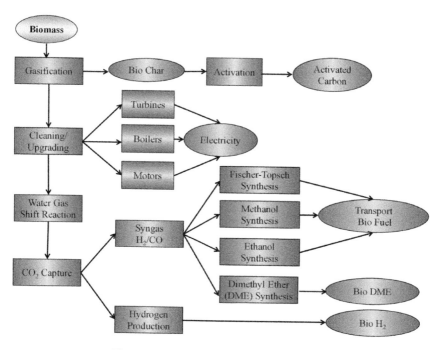

Figure 5.2. Biomass gasification main products.

of biomass, forming light gases, tar, and char. Gases contain mainly H_2, CO, CO_2, H_2O, NH_3 (ammonia), H_2S (hydrogen sulphide) phenols, CH_4, and other hydrocarbons. Tar fraction contains heavier organic compounds that are gaseous when they are released, but at normal conditions of pressure and temperature are doughy (on the border between solids liquids). Char is composed mainly from carbon, hydrogen, and the mineral matter present on the solid biomass. Afterwards, at even higher temperature, occur char gasification, char reacts with surrounding atmosphere, including gasification agent and volatiles released during the devolatilisation stage. A large number of homogeneous reactions occur also between all the species present on the gas phase. Cracking reactions causes a breakage of the heavier hydrocarbon molecules, with the formation of lower mass gaseous species. In Table 5.1 are presented the main gasification reactions.

Table 5.1. Biomass main gasification reactions.

Designation	Reaction	ΔH (kJ/mol)	
Oxidation	$C_{(s)} + O_2 \leftrightarrows CO_2$	−392,5	(1)
	$C_{(s)} + \frac{1}{2} O_2 \leftrightarrows CO$	−110,5	(2)
	$CO + \frac{1}{2} O_2 \leftrightarrows CO_2$	−282	(3)
Boudouard	$C_{(s)} + CO_2 \leftrightarrows 2\,CO$	172,0	(4)
Water Gas: primary	$C_{(s)} + H_2O \leftrightarrows CO + H_2$	131,4	(5)
secondary	$C_{(s)} + 2\,H_2O \leftrightarrows CO_2 + 2\,H_2$	90,4	(6)
	$C_{(s)} + 2\,H_2 \leftrightarrows CH_4$	−74,6[†]	(7)
Methanation	$CO + 3H_2 \Leftrightarrow CH_4 + H_2O$		(8)
	$CO_2 + 4H_2 \Leftrightarrow CH_4 + 2H_2O$		(9)
Water-gas shift	$CO + H_2O \leftrightarrows CO_2 + H_2$	−41,0	(10)
Steam Reforming	$CH_4 + H_2O \leftrightarrows CO + 3\,H_2$	205,9[†]	(11)
	$CH_4 + 2\,H_2O \leftrightarrows CO_2 + 4\,H_2$	164,7[†]	(12)
	$C_nH_m + n\,H_2O \leftrightarrows n\,CO + (n + m/2)\,H_2$	210,1[†‡]	(13)
	$C_nH_m + n/2\,H_2O \leftrightarrows n/2\,CO + (m-n)\,H_2 + n/2\,CH_4$	4,2[†‡]	(14)
CO$_2$ Reforming	$CH_4 + CO_2 \leftrightarrows 2\,CO + 2\,H_2$	247,0[†]	(15)
	$C_nH_m + n\,CO_2 \leftrightarrows 2n\,CO + m/2\,H_2$	292,4[†‡]	(16)
	$C_nH_m + n/4\,CO_2 \leftrightarrows n/2\,CO + (m-3n/2)\,H_2 + (3n/4)\,CH_4$	45,3[†‡]	(17)
H$_2$ Reforming	$CO + 3\,H_2 \leftrightarrows CH_4 + H_2O$	−205,9[†]	(18)
Hydrogenation	$CO + 2H_2 \Leftrightarrow CH_4 + \frac{1}{2} O_2$		(19)
Cracking	$p\,C_nH_m \leftrightarrows q\,C_xH_y + C_zH_u + r\,H_2$ (x, z < n and y, u < m)		(20)

The yield and composition of the gas produced depends on gasification agent. Usually oxygen (or air) and steam are used as gasification agent, but CO_2 or mixtures of any of these components may also be used. The presence of oxygen (or air) is advisable, as they promote partial combustion of the feedstock, and hence the release of the energy needed for gasification. When air is used, the gas produced is diluted with nitrogen, thus it presents a lower calorific value, not suitable for some application, namely chemical synthesis. The use of oxygen instead of air solves the problem of nitrogen dilution, but due to the cost of oxygen production, operative costs increases. The choice of gasification agent depends on the application of gasification gas, as shown in Fig. 5.2. Lately, research and development of thermochemical technologies has been focusing on the production of alcohol mixtures and DME by chemical syntheses, instead of diesel like biofuels by Fischer-Tropsch synthesis, mainly due to economic reasons and absence of biomass supply chain at large scale and at competitive prices.

There is not much information about algal biomass gasification, because being a thermochemical process at relatively high temperature, gasification requires low moisture content, as mentioned before. Thus, little published information is available. Hirano et al. 1998 gasified *Spirulina* at temperature from 850°C to 1000°C, and reported that the methanol yield was 0.64 g per 1 g of biomass. The low energy balance was due to the intensive energy needed for algae harvesting during centrifuge process. Minowa and Sawayama (1999) gasified the microalgae *Chlorella vulgaris* to produce methane-rich fuel. One possible way of overcoming the problems related to the use of algal biomass in gasification processes is to co-gasify algal biomass mixed with other biomass types with lower water content.

2.3 Pyrolysis

Nowadays, the main objective of biomass pyrolysis is the production of products that are liquid at normal conditions of temperature and pressure. Biomass pyrolysis occurs in the absence of air at temperature in the range from 450°C to 550°C and at atmospheric pressure. To maximise the production of liquids, high heating rates and short residence times are needed. These liquids can be used as raw material in several industries or as biofuel after being refined to be converted into secondary fuels, more valuable and suitable to be used in the transport sector. Another option is the production of bio chemicals, as shown in Fig. 5.3.

Biomass pyrolysis also produces gases (syngas) and biochar, though in much smaller quantities than liquids. The gases can be used as fuel and the major constituents are: H_2, CO, CO_2, N_2, and gaseous hydrocarbons. After activation biochar may be converted into activated carbon. The yields of these main products: syngas, liquid biofuel and biochar depends on the

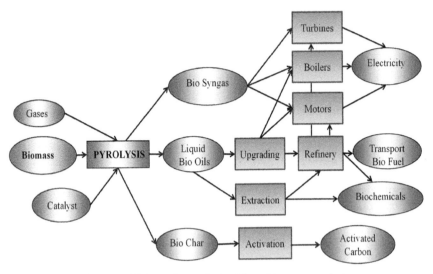

Figure 5.3. Main products obtained from biomass pyrolysis.

technology used, type of reactor and of operational conditions: type of gas, pressure, temperature, reaction time, type of solvent, and catalyst. The type and composition of biomass is also very importance.

To increase the production of liquids, it is fundamental to use very high heating rates, thus fine particle size of biomass are needed to achieve rapid heat transfer and very low residence times. Liquid molecules once formed, need to be removed from the reaction medium and cooled down to prevent their further reaction and the conversion of the liquids initial formed into lighter molecules, gaseous at normal temperature and pressure conditions, or the conversion into heavier molecules, solid at normal temperature and pressure conditions, as shown in Fig. 5.4. This process is referred to as flash pyrolysis, and it allows obtaining liquids yields of around 75%, while gas and solid yields are around 12%. Fast pyrolysis uses slightly lower heating rates and not so low residence times, which is reflected in lower yields of liquids, around 50%, whilst gas and solid yields are around 25% each. The main products of slow pyrolysis are char and gases, with yields that may reach values around 45 and 35%, respectively, while liquids are usually below 20%. Biomass fast and flash pyrolysis have been studied by many authors (for instance Yildiz et al. 2016, Bridgwater and Peacocke 2000, Bridgwater et al. 1999).

Usually bio oils obtained by biomass pyrolysis present several disadvantageous characteristics and properties, namely very high oxygen content, in the rage of 30 to 40%, low pH values, usually between 2.5 and 4.0, and lower heating value (16–20 MJ/kg) than that found in fossil fuel

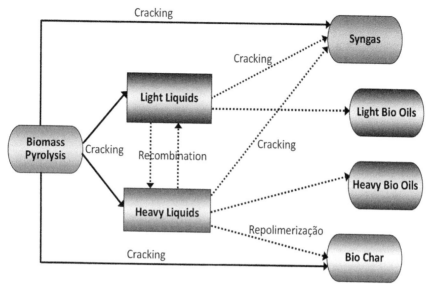

Figure 5.4. Possible mechanisms for biomass pyrolysis. First reactions are presented as straight lines and secondary reactions are shown as dotted lines.

oils. Most of the oxygen appears as water, thus its contents are between 15 and 30%. These bio oils are also chemically unstable, and its physical and chemical properties change during storage. Other disadvantages of pyrolysis bio oils are: high viscosity, low volatility, corrosiveness, and cold flow problems. Thus, to improve pyrolysis bio oils properties some upgrading and refining is needed.

Fast and flash pyrolysis technologies were developed on a commercial scale and some pilot plant with fluidised bed reactors mainly for forestry biomass were built in USA, Canada, and Europe.

As microalgae present high content of cellular lipids, resolvable polysaccharides, and proteins, in comparison to lignocellulose, they may be converted into bio oils and bio-gases by pyrolysis, the main problem is the great deal of energy needed to remove moisture in the algal cells. However, pyrolysis of algal biomass research work is more extensive than gasification. Some authors have studied pyrolysis of some algae species, namely: *Chlorella, Emiliania huxleyi, Nannochloropsis* sp., *Plocamium, Sargassum, Spirulina/Arthrospira, Synechococcus, Tetraselmis,* and cultivated mixed consortia. Huang et al. 2010 and Saber et al. 2016 reviewed the works published by some authors, as can be seen in Table 5.2. The high yields of liquids are obtained because not only the lipids, but also protein and water-soluble carbohydrate are converted into oil yields. Yang et al. 2016

Table 5.2. Different types of catalysts performance SCWG of algal biomass (adapted from Patel et al. 2016 and Saber et al. 2016).

Species	Type of Pyrolysis	Conditions	Oil yield (%)	Reference
Auxenochlorella protothecoides (formerly *Chlorella protothecoides*)	Fast	500°C, 600°C/s	52.0[a]	Peng and Wu 2000
A. protothecoides (formerly *C. protothecoides*)	Fast	500°C, 600°C/s	18[a]	Miao and Wu 2004a
Microcystis aeruginosa	Fast	500°C, 600°C/s	24[a]	Miao and Wu 2004a
Heterotrophic *A. protothecoides* (formerly *C. protothecoides*)	Fast	450°C, 600°C/s	57.9[a]	Miao and Wu 2004b
A. protothecoides (formerly *C. protothecoides*)	Fast	500°C, 10 K/s	53.3[a]	Demirbas et al. 2006
Tetraselmis chui, *Chlorella* like, *Chlorella vulgaris*, *Chaetoceros muelleri*, *Dunaliella tertiolecta*, *Synechococcus*	Slow	700°C, 10°C/min	43	Miao and Wu 2004b
Nannochloropsis sp. residue	Slow	300–500°C, 10°C/min	31.1	Pan et al. 2010
Blue-green algae blooms (BGAB)	Slow	300–700°C, 15°C/min	54.9	Hu et al. 2013
Chlorella vulgaris remnants	Fast	500°C	53	Wang et al. 2013
Macroalgae (seaweeds)	Microwave	200–300 W	21	Budarin et al. 2011
Chlorella sp.	Microwave	500–1250 W, 462–627°C, 20 min	28.6	Du et al. 2011

state that dry algal biomass can be converted into biocrude, charcoal, and a gas by pyrolysis at atmospheric pressure and at a temperature between 400–600°C, as the biocrude presents high viscosity and water content, and some upgrading is needed, including dewatering and hydrotreatment to remove oxygen, sulphur, and nitrogen. However, Yang et al. 2016 stated that hydrothermal liquefaction is more advantageous to transform algae into biofuel, because dewatering is not needed.

To improve bio oils quality and yield, several catalysts have been used, such as: Co/Al$_2$O$_3$, Ni/Al$_2$O$_3$, γ-Al$_2$O$_3$, ZSM-5, HZSM-5, and nickel phosphide (Saber et al. 2016).

Hu et al. 2013 used a different process, microwave-assisted pyrolysis, for converting *Chlorella vulgaris* into bio oils. The results obtained showed that at the microwave power of 1500 W, a bio oil yield of 35.8 wt.% was reached. It was also reported that in presence of catalyst, Pyrolysis of *C. vulgaris* was improved, of the catalysts tested, activated carbon was the one that led to the best results.

3. New Hydrothermal or Wet Processes

Algae cell main compounds are lipids, carbohydrates, and proteins, which can be converted into biofuels and/or bio-products (Patel et al. 2016). However, as algae grow in diluted water media, the amount of energy needed to concentrate algae to be processed is quite high. On the other hand, conventional thermal processes requires low water contents and thus intensive thermal drying processes are essential, which makes the overall process prohibitive (Elliott 2016). New hydrothermal processes are suitable to deal with feedstocks with high water contents, generally up to 70 wt.% or more. Thus, these processes show important advantages towards conventional ones, as they do not require intense drying processes. These new processes also have some advantages in relation to anaerobic digestion methods, because while these later processes need 2 to 4 weeks, hydrothermal processes only require a few minutes and much smaller equipment (Yakaboylu et al. 2015).

Hydrothermal processes may be divided into: hydrothermal carbonization (HTC), hydrothermal liquefaction (HTL), and supercritical water gasification (SCWG), as presented in Fig. 5.1, together with the main experimental conditions. The main product of HTC is a solid usually referred as biochar and the temperature used range from 180°C to 250°C (Castello et al. 2014). In HTL, sub-critical conditions are used and temperature varies between 250°C and 373°C. The main products are liquids, though some gases may also be formed (Elliott et al. 2015). When temperature and pressure are above the critical values for water (373.95°C and 22.06 MPa), the main products are gaseous and the process is called SCWG (Castello et al. 2014). At subcritical conditions (200–370°C and 4–22 MPa), water has a different behaviour from water at room temperature and also from supercritical water, as shown in Fig. 5.5. As water has shown supercritical characteristics at temperatures even below 300°C, it has been used as reaction medium at subcritical conditions. At subcritical conditions, water shows properties of a non-polar solvent, but the structure of each molecule remains unaffected, thus it behaves as a polar molecule that can interact, for instance with ions (Prado et al. 2016). Many authors have studied the application of these hydrothermal or wet processes to different types of biomass, including algae, as will be discussed next.

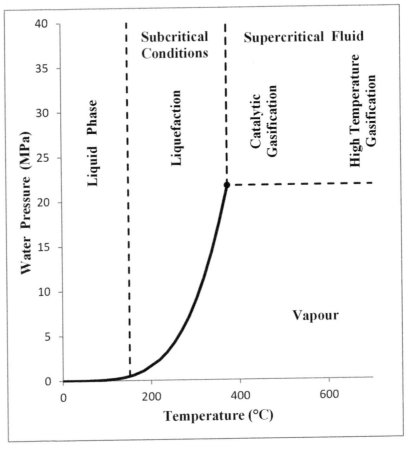

Figure 5.5. Water pressure versus temperature diagram.

3.1 Supercritical Water Gasification

3.1.1 Introduction

SCWG is a type of gasification, whose main compounds are also CO, CO_2, H_2, and CH_4. The main differences between SCWG and conventional gasification is that the latter process uses supercritical water conditions (374°C and 22.1 MPa) and water acts as a reactant and as gasification medium. At supercritical conditions, the formation of H^+ and of OH^- has a great probability, which promotes hydrolysis and pyrolysis reactions, the formation of free radicals is also greatly promoted, which favours biomass components degradation (Reddy et al. 2014).

As reported by Yakaboylu et al. 2015, when biomass moisture content is higher than 30%, SCWG is energetically more favourable than other gasification technologies. SCWG is a clean and fast process that can decompose cellulose, hemi-cellulose, and lignin. It has several advantages: pre-treatments are not needed, shorter reaction time, lower residue formation, and lower generation of degradation products, no use of toxic solvents and less corrosion (Prado et al. 2016). The main drawback is the lack of full development and the absence of industrial scale installations.

First studies of SCWG of biomass were done with modular simple compounds, including sugars and guaiacol to study the behaviour of cellulose, hemicellulose, and lignin-derived compounds. Cellulose and hemicellulose dissolution is favoured by the high temperature and pressure used in SCWG, leading to the formation of simple sugars with 5 and 6 carbon atoms, which are next converted into acids such as: carboxylic, succinic, acetic, etc. Lignin is converted into phenolic compounds, such as guaiacols and syringols, aromatics and aldehydes. Alcohols like coumaryl, coniferyl, sinapyl are also formed. Afterwards these compounds are converted into CO, CO_2, H_2, and CH_4 (Reddy et al. 2014). Some of the reactions mentioned before: steam reforming, methanation, hydrogenation, and water gas shift reactions may also occur. The formation of C_2–C_4 hydrocarbons may also happen, depending on biomass composition and SCWG conditions. In Fig. 5.6, a simplified diagram about main biomass constituents (cellulose, hemicellulose, and lignin) conversion into gases under SCWG conditions is presented.

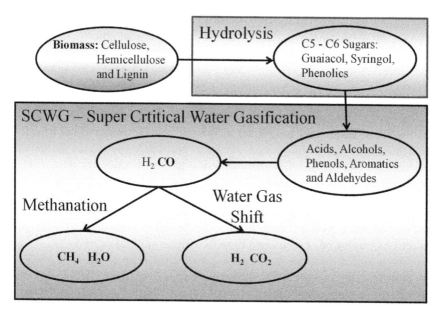

Figure 5.6. Conversion of main biomass constituents (cellulose, hemicellulose, and lignin) into gases under SCWG conditions.

The number of reactions that happen during biomass conversion into gases under super critical water conditions is huge, however, the main reactions that occur during SCWG may be summarised by reactions (21) to (26), besides these ones, methanation reactions (8) and (9), water gas shift reaction (10), and hydrogenation reaction (19) also occur (Reddy et al. 2014):

$$CH_xO_y + (2\text{-}y)\ H_2O \Leftrightarrow CO_2 + (2\text{-}y+x/2)\ H_2 \tag{21}$$

$$CH_xO_y + (1\text{-}y)\ H_2O \Leftrightarrow CO + (2\text{-}y+x/2)\ H_2 \tag{22}$$

Cellulose hydrolysis

$$C_6H_{10}O_5 + H_2O \Leftrightarrow C_6H_{12}O_6 \tag{23}$$

Glucose reforming reaction

$$C_6H_{12}O_6 \Leftrightarrow 6\ CO + 6\ H_2 \tag{24}$$

Hydrolysis of lignin

$$C_{10}H_{10}O_3 + H_2O \Leftrightarrow C_{10}H_{12}O_4 \Leftrightarrow \text{Phenolic Compounds} \tag{25}$$

Steam reforming reaction

$$\text{Phenolic Compounds} + H_2O \Leftrightarrow CO + CO_2 + H_2 \tag{26}$$

3.1.2 Effect of SCWG conditions

Temperature is one of the most important parameters in SCWG, because the conversion of biomass complex molecules needs a great amount of energy. Temperatures in the range from 350°C to 500°C favour the formation of CH_4 and CO, as methanation reactions are promoted, while temperatures between 500°C and 800°C promotes the formation of H_2 and CO_2, as reforming and water gas shift reactions are favoured.

Another important parameter is residence time defined as the time in which reactants stay in contact at reaction conditions. SCWG usually requires residence times of some seconds. Normally the rise of residence time led to higher gasification efficiency and to higher content of H_2. However, after the upper limit is achieved, further increases of residence time lead to insignificant changes. The best residence time is much affected by the type of biomass and by the other conditions used for SCWG, for instance at 390°C to 450°C and 25 MPa, the de-polymerization of lignin can be achieved after 5s of residence time (Reddy et al. 2014).

The effect of pressure is quite complex. At the critical point, 22.1 MPa, the ionic reaction of water is favoured in detriment of free radical reactions, because of the high densities, a solvent cage around the solute molecules is formed, thus reactions between solutes like coking and polymerization are restrained, while reactions between solute and solvent, like reforming and water gas shift reactions are favoured. The effect of pressure is much affected by temperature, being more pronounced at higher temperatures. It was reported that the rise of pressure favoured the decomposition rate of lignin and the yields of H_2, CH_4, and CO_2 till a certain value of pressure was attained at 400°C, afterwards no significant changes were observed (Reddy et al. 2014). Again the effect of pressure was dependent on biomass type and other SCWG conditions. For instance, gasification efficiency of lignin was enhanced and higher yields of H_2 and CH_4 and lower CO_2 yields were obtained with the rise of pressure from 15 to 27.5 MPa. On the other hand, for glucose gasification the yields of H_2 decreased with the rise in pressure at temperatures between 400° and 600°C.

Another important parameter is biomass concentrations, as water may have a dual role of catalysts and solvent, its presence is most important. In general the increase of biomass concentration leads to lower biomass conversion into gases. As reported by Reddy et al. 2014, the increase of glucose concentration from 1 to 17% led to a decrease of H_2, CH_4, and CO_2 yields. The rise of lignin concentration from 5 to 33 wt.% at 600°C favoured the release of CH_4, but decreased the yields of H_2 and CO. The increase in glycerol concentration improved CH_4 yields at the expense of H_2. The best biomass concentration is also dependent on biomass type and SCWG conditions.

3.1.3 Catalysts used in SCWG

The presence of suitable catalysts may improve considerable biomass SCWG. Reddy et al. 2014 reviewed the main important catalysts used for SCWG of different biomass feedstocks, namely alkali metals, transition metals, and different activated carbons. Alkali metal such as: KOH, NaOH, Na_2CO_3, $KHCO_3$, and K_2CO_3, have been reported to be effective catalysts for biomass SCWG, by improving SCWG reactions, namely water gas shift, and also by favouring H_2 release. These homogeneous catalyst are usually dissolved in super critical water, which makes difficult its recuperation, another drawbacks are reactor plugging, corrosion, and fouling.

Several transition metals: Ni, Ru, Cu, and Co in various supports (γ-Al_2O_3, ZrO_2 and activated carbon) and with different promoters (Na, K, Mg, or Ru) have been investigated for SCWG. In general, transition metals have high catalytic activity for SCWG reactions. These heterogeneous catalysts usually present high catalytic activity and selectivity. Ni-based catalysts enhance the release of H_2 and CH_4 and are cost-effective, however, they may lead to sintering under SCWG conditions (Reddy et al. 2014). Ru-based catalysts also have high selectivity towards H_2, and also have the advantage of resisting more to sintering than Ni-based catalysts. Reddy et al. 2014 reported that Ru, Cu, and Ag doped with Ni showed better catalytic activity and stability. Ru supported on rutile TiO_2 and carbon also showed high stability under SCWG conditions. On the other hand, the char formed during SCWG usually deposits on the surface of transition metals catalysts, which decreases the catalytic activity. The structure of the supports used for transition metals catalysts are usually unstable, and the changes observed reduces the catalytic activity. Transition metals catalysts may be recovered by filtration followed by drying. Ru is normally easier to recover than Rh, Ni, and Pt, and with lower recovery costs.

Activated carbons like spruce wood charcoal, macadamia shell charcoal, coal activated carbon have been tested as SCWG catalysts. In general, they are a good option to catalyse biomass SCWG, by enhancing WGS and methanation reactions and improving gasification efficiency. However, activated carbons performance is enhanced when they are impregnated with metals. Lignin gasification efficiency improves by the order: Ru/C > Rh/C > Pt/C > Pd/C > Ni/C, whilst H_2 yield follows the order: Pd/C > Ru/C > Pt/C > Rh/C > Ni/C (Reddy et al. 2014). The surface area of activated carbon catalyst may be decreased due to chemisorption of the intermediates, which reduces catalyst activity. Activated carbon catalysts may be recovered by acid washing of the remaining solid products.

3.1.4 Types of reactors used for SCWG

SCWG studies have been done using batch, mainly autoclaves, and continuous reactors, like: tubular reactors, continuous stirred tank reactors, and fluidized bed reactors. Reddy et al. 2014 presented a summary of the main works about SCWG in batch and in continuous reactors, using different types of feedstocks. In the discontinuous SCWG reactors tests, model compounds were used (glucose, cellulose, lignin, hemicellulose), and also sawdust, sea weed, hog manure, and sugar cane bagasse. In the continuous SCWG reactors, studies with glucose, organic waste, wheat straw, pulping black liquor, cauliflower residue, tomatoes residue, acorn, extracted acorn, and hazelnut shell have been reported. Due to the high needs of energy, continuous reactors are a best option for SCWG, especially tubular reactors, because of the low residence times required for SCWG. Some continuous stirred tank reactors have also been used, that are a combination of autoclave and tubular reactor, but the higher energy demand and complexity are important disadvantages of this type of reactors. Fluidized bed reactors are also a good option for SCWG, due to the conditions used, high temperatures and pressures, a new correlation for minimum fluidization velocity has to be developed. Lu et al. 2013 studied the hydrodynamics of SCWG in fluidized bed reactors and proposed a correlation for minimum fluidization velocity for temperatures from 360°C and 420°C, and for pressures in the range of 23 to 27 MPa. Continuous-flow microchannel reactors are a promising option, as the high heat transfer rates needed for SCWG reactions may be easily achieved. High gasification conversions and high H_2 yields were reported for this new type of reactor (Goodwin and Rorrer 2009).

3.1.5 SCWG of algal biomass

Biomass model compounds for cellulose, hemi cellulose, and lignin have used to study and understand the complex reactions that happen during SCWG. As mentioned before, SCWG has also been applied to different types of biomass: algal biomass, cattle manure, sewage sludge, lignocellulosic biomasses, fruit wastes, food wastes, paper waste sludge, sugarcane bagasse, etc. Different types of microalgae and macro algae have been investigated. Products yields and properties depend on feedstock composition, including cell wall composition and structure, the type of monosaccharides present and the type of bonds among them and lignin (Prado et al. 2016). Thus, the optimisation of process conditions is affected by the type of algal biomass used in SCWG.

Besides, the feedstock composition, algal biomass conversion by SCWG is affected by the following parameters: reactor design, time and reaction temperature, and reactant medium, including the presence of catalysts, as mentioned before for SCWG of general biomass. Pressure is usually considered as non-key parameter, however, as temperature is an important parameter to achieve fast reactions, when temperatures is above water boiling point, pressure has to be high enough to keep water in the liquid phase and to avoid phase transition, as was seen in Fig. 5.5 (Brunner 2009, Cantero et al. 2015b, Schacht et al. 2008). However, Cantero et al. 2015a,b stated that degradation of cellulose and glucose and the yield of 5-HMF (5-hydroxymethylfurfural) increased when pressure increased at constant temperature. Öztürk et al. 2010 also reported that pressure affected the degradation rate when working at subcritical water.

SCWG converts algae into gaseous compounds, whose main components are: CO, H_2, CO_2, CH_4, and C_2–C_4 hydrocarbons. Several reactions occur during algae SCWG: hydrolytic conversion of macro-molecules into smaller organics, gasification of these organics, forming CO, CO_2, and H_2, water gas shift reaction of CO, by reaction (10) forming CO_2 and H_2 and methanation of CO and CO_2 by reaction (8) and (9) (Tiong et al. 2016). Water is a real reactant of SCWG, and several authors have reported the decrease of water after experiments. Graz et al. 2016 reported that after SCWG tests, the liquid residue was around 77 ± 3 wt.% of the initial water.

Due to the presence of water in SCWG, lower tar contents are obtained, together with higher efficiency and the possibility of using feedstocks with high N contents to obtain a nitrogen free fuel, in comparison with conventional gasification. The best SCWG conditions are difficult to predict based on the information available for terrestrial biomass, due to the high range of different algal composition. Temperature and residence time are important parameters on gaseous product distribution. For instance, temperatures up to 500°C enhance the production of CH_4, while higher values favour the formation of H_2. On the other hand, the rise of residence time promotes the formation of CH_4 and H_2.

Graz et al. 2016 also studied the effect of the main experimental conditions on SCWG of *Ulva rigida* (formely *Ulva armoricana*) and *Ulva rotundata* macroalgae, namely: temperatures from 400°C to 550°C, reaction time in the range of 7 and 120 min and algae concentrations between 7 and 16.4 wt.%. The results obtained showed that the lowest reaction time tested, 7 min, was enough to obtain a significant conversion, however, longer reaction times increased the formation of CH_4 through reaction (24). The gas obtained with the reaction time of 120 min led to a CH_4/H_2 ratio of 2.5. Gas composition was highly affected by temperature. Graz et al. 2016 reported that an increase in temperature from 400 to 550°C led to an increase in H_2 contents from 3 to about 16 mole %. On the other hand, CH_4 content increased till 500°C, and afterwards decreased till 12 mol.% at 550°C, because the rise of temperature promoted reaction (25) and inhibited reaction (26). On the other hand, no significant changes were obtained in gas yield when the amount of macro algae increased from 7 and 16.4 wt.%. However, H_2 and CH_4 contents decreased around 30% each, but CH_4/H_2 ratio remained around 0.8.

Ekpo et al. 2016 compared SCWG of microalgae *Chlorella vulgaris* with SCWG of other wet biomasses, namely: sewage sludge digestate, swine and chicken manure and found that at 500°C the highest gas yield was obtained for microalgae with values of 55 wt.%, while the other feedstocks presented values around 30 wt.%. Microalgae also led to the lowest yield of residues ranged from 10 to 15 wt.%. These results agree with those reported by Yakaboylu et al. 2015 that studied SCWG of microalgae, focusing on the effect of the inorganic content of the biomass, on the energetic performance, and on the behaviour of formed compounds. These authors' simulations used temperatures of 400°C, 500°C, and 600°C, considering dry microalgae concentrations from 10 to 30 wt.%. The results obtained showed that the rise of the mass fraction of microalgae inorganic content favoured the formation of CH_4 and H_2 and reduced the need of thermal energy for the reactor. Lower temperatures also increased the thermally efficiency of the process. However, the high energy need of the SCWG is a drawback of the process, but part of the thermal energy can be recovered and used for microalgae growth (Yakaboylu et al. 2015).

Tiong et al. 2016 studied the influence of microalgae type on gas composition obtained by SCWG, using *Chlorella vulgaris*, which has high protein and low carbohydrate contents and *Scenedesmus quadricauda* with low protein and high carbohydrate contents. In general, both species presented similar total gas yield both in absence and in presence of nickel based catalysts, though the use of catalyst allowed increasing gas yields to 80 to 90%. Without catalyst the main gas obtained was CO_2, probably formed by decarboxylation reactions, while the presence of catalysts led to several gaseous components by the order $CH_4 > CO_2 > H_2 > CO$.

Algae conversion through SCWG is a promising option, however, to overcome some problems of this technology, its combination with other hydrothermal processes (namely SCWG of HTC solids or of HTL liquids) are probably good options, though the overall costs need to be calculated and analysed.

Some authors have studied the integration of several hydrothermal processes to increase products yields and quality and to achieve better thermal efficiencies and energy recovery. Cherad et al. 2016 studied the integration of HTL with SCWG for *Chlorella vulgaris* valorisation. HTL at 350°C and during 60 min led to biocrude yields around 30%, however reaction time had only a small effect in biocrude yield, as an increase of around 3% was observed from 30 to 60 min. The aqueous phase obtained by HTL was upgraded by SCWG in presence of sodium hydroxide. High yields of H_2 (30 mole H_2/kg algae) and organics gasification of around 98% were reported. Final aqueous phase was still rich in nutrients that can be recycled for algal growth.

3.1.6 Catalytic SCWG of algal biomass

To improve algae conversion by SCWG, different catalysts have been tested. Patel et al. 2016 summarised the most used catalysts tested for SCWG of algal biomass. Different catalysts have been tested. The same types of catalysts tested for general biomass were also used for algal biomass: alkali metals, transition metals, and activated carbons. The use of KOH and NaOH allowed achieving a cold gas efficiency of around 92%, and led to an increase in the heating value of the gas produced, this gas is poor in CO_2. The production of both CH_4 and H_2 are favoured by the presence of Ru based catalyst. As mentioned before for general biomass, the best results are achieved when heterogeneous catalysts are used for algal biomass SCWG. Patel et al. 2016 reported that the combination of homogeneous catalyst with Ru/Al_2O_3 allowed increasing cold gas efficiency to around 109%. Ru-based catalysts showed to be a best option than other transition metals. According to Chakinala et al. (2009), the production of H_2 in presence of transition metals catalyst usually follows the order: Ru > NiMo > Inconel > Ni > PtPd

Table 5.3. Different types of catalysts performance SCWG of algal biomass (adapted from Patel et al. 2016).

Catalyst	Temperature (°C)	Time (min)	Feed Feedstock (wt. %)	CGD (%)	Reference
0.5M, 1.5M, 1.67M, 3M NaOH	500		0.33–13	29–93	Cherad et al. 2013, 2014
5% Ni/Al$_2$O$_3$	500	30	5–6.7	65–77	
5%, 10%, 20% Ru/ Al$_2$O$_3$	500	0–120	3–13	60–97	
2% Ru/ZrO$_2$	400	61–360	2.5–20	18–93	Stucki et al. 2009
2% Ru/TiO$_2$	400–700	1–15	2.9–7.3	29–70	Chakinala et al. 2009
Pt Pd/Al$_2$O$_3$	400–700	1–15	2.9–7.3	21–70	

CGD–cold gas efficiency

> CoMo. Cherad et al. 2014, 2013 reported that the use of Ru/Al$_2$O$_3$ led to increases in H$_2$ and CH$_4$ yields of about, 100% and 200%, respectively.

The presence of inorganic salts in algae may act as catalyst for SCWG, but they may also lead to the formation of different salts that poison catalysts and decreases their life time and their action. The same effect is caused by H$_2$S, whose formation may happen even when low sulphur contents are found in algae.

In Table 5.3, the effect of using different types of catalysts on cold gas efficiency obtained for SCWG of algal biomass are presented (Patel et al. 2016).

To overcome the problems that may arise due to the presence of inorganic salts in algal biomass, Stucki et al. 2009 suggested a two stage process, in the first one the inorganic salts were removed, and in the second stage the suitable catalyst would be used to promote the SCWG process.

3.2 Hydrothermal Liquefaction

Hydrothermal liquefaction (HTL) could be defined as a thermochemical conversion process in water medium at low/medium temperatures (280 to 374°C) and high pressure (4 to 22 MPa) conditions (Douglas et al. 2015). At near-critical conditions, water becomes highly reactive due to drastic changes in properties such as solubility, density, dielectric constant, and ion product. Under these conditions, hot compressed water becomes highly attractive for many chemical reactions that would not occur at standard water conditions. As the miscibility of algae compounds (mainly composed of lipids, carbohydrates, and proteins) significantly increase at near critical conditions, chemistry in a single phase is favourable when compared to a multi-phase medium (solid, liquid and vapour phases). As a result, algae conversion through HTL outcomes for a stable liquid-phase product (bio-oil) and other products in gaseous, aqueous, and solid phases (Barreiro et al. 2013).

The terms hydrothermal liquefaction, liquefaction, and hydro-liquefaction are synonymous for processes in which wet biomass is converted into bio-oil through specific temperature and pressure conditions. A different definition (pressurized aqueous pyrolysis) was proposed by other researchers (Marcilla et al. 2013), although a number of differences were found in bio-oil characteristics mostly in regard to lower oxygen and moisture content when compared with conventional pyrolysis process.

HTL presents a viable route for converting a wide range of materials (including wastes) into liquid products of combustible nature and chemicals without the need of pre-drying. For fuel purposes, these bio oils require additional upgrading to fine-tune a number of physical and chemical properties through different processes (e.g., solvent extraction, distillation, hydro-deoxygenation, and catalytic cracking). Those properties that need to be changed and adjusted, in order to meet current liquid fuel standards, include viscosity, density, heating value, oxygen, nitrogen and sulphur content, and chemical composition.

The conversion of natural solid biomass into liquid fuels by HTL requires specific conditions. Biomass is characterized as being a non-homogeneous material (in its physical form) with lower energy density and higher moisture content. At HTL conditions in water medium, biomass macromolecules are breakdown (including carbohydrates, lignin, lipids and proteins) into lower molecular weight structures. As the HTL process continues, those lower molecular weight structures could be continuously broken into even smaller fragments.

In the last 15 to 20 years, world biofuel production has been mainly focused on first-generation fuels towards the production of ethanol and biodiesel from different sources (e.g., starch, sugars, and vegetable oils). The production based on advanced biofuels or biofuels produced from lignocellulosic materials, such as wood waste and straw, only contributed to 0.2% of total biofuel production (Nakada et al. 2014). These issues have been addressed by a number of processing technologies towards its use mostly in internal combustion and aviation engines. Only recently, significant developments on research were achieved on using biomass as source of liquid fuels.

A great number of publications may be found in literature based on pyrolysis and hydrothermal liquefaction of lignocellulosic materials and upgrading processes. Also, a number of technologies have been currently employed with the aim of converting biomass residues into valuable-added products. Among these, thermochemical conversion processes are of significant importance, due to their ability to convert biomasses into different types of liquid biofuels and chemicals with increased handling, distribution and storage capability.

When compared with other thermochemical processes, HTL presents some advantages in the conversion of algae biomass into biofuels mostly in regard to feedstocks characteristics (e.g., high level of moisture content and heterogeneity) and its integration with other processes. However, issues such as location of cultivation, species selection, amount and quality of biofuel and by-products produced, and its sustainability are critical when the implementation of a large-scale industrial facility is of concern. In addition, bio-oil upgrading is still required to meet current petroleum standards or to produce similar liquid fractions for integration in common refinery processes.

3.2.1 Process nature

The nature of the algae HTL process could be chosen over other thermochemical processes, as no water removal from feedstock is required through pre-drying prior to its conversion into a high energy density bio-oil (Jena and Das 2011). During the conversion of long carbon chain molecules into smaller ones, oxygen is removed in the form of H_2O (dehydration) and CO_2 (decarboxylation) which result in a bio-oil with high H/C ratio (Goudriaan and Peferoen 1990). At near-critical and supercritical conditions, water behaves more like a non-polar solvent with strong catalyst effect. As a result, whereas organic substances are insoluble in water under normal conditions, water acts as a good solvent for non-polar substances under these conditions.

Taking into account the main characteristics of HTL process, different feedstocks sources could be used to produce bio oils with fuel-potential properties. Over the past three decades, a significant number of research have been published in regard to economic valorisation of different biomass wastes including agricultural (Demirbaş 2000, Demirbaş 2005, Mazaheri et al. 2010), forest (Araya et al. 1986, Minowa et al. 1998), blending materials (Karagöz et al. 2005) and algae (Dote et al. 1994, Minowa et al. 1995, Toor et al. 2013, López et al. 2013, Neveux et al. 2014). Comparing the research findings, significant effort is required to define the base-criteria for the best/optimal algae HTL due to the number of different species used, their initial state, the operation conditions and even the catalyst used.

Different homogeneous and heterogeneous catalysts may be included to improve the quality of bio-products and yields. The use of catalyst has been reported in literature in order to enhance or inhibit the formation of different products. Such examples may be found when alkali carbonates catalysts are used to enhance the formation of liquids and inhibit the formation of char and gaseous products, while nickel catalysts tend to enhance the formation of gases. Catalysts used may include water-soluble inorganic compounds and salts (KOH and Na_2CO_3), organic acids (acetic and formic acids), and transition metal catalysts (Ni, Pd, Pt and Ru) supported on carbon, silica or alumina base. Both reaction medium and catalyst are of great importance regarding reduction of mass transfer resistances and of fragmentation of biomass internal structure (Peterson et al. 2008, Chumpoo and Prasassarakich 2010). Until now, with different types of catalyst to improve conversion ratios, algae HTL could lead to 60 wt.% of bio oils yields.

Although much has been published on the subject of HTL, most of the research is focusing on small-scale type reactors. While feasible at bench-scale, to date, the setup to industrial-scale facility based on algae HTL is quite limited as well as contradictory. For a large-scale production, reactors are usually designed accordingly to type of reactions (either endothermic or exothermic) as significant differences are found in regard to heat transfer systems (supply or removal of heat). Moreover, the energy needed for HTL reactions has not yet been fully determined for different algae sources, which is fundamental for the design of large-scale commercial continuous-flow reactors with a proper heat transfer system. Available data seems to be rather limited and sometimes inconsistent, as overall reaction has been reported as being either endothermic or mildly exothermic (Lee et al. 2016).

The production of biofuels through HTL is still a new process and a number of issues, namely type of feedstock and complex feed systems, higher investment costs, and low value of final product, still need to be carefully analysed (Marcilla et al. 2013). However, this process is currently attracting much interest as the conversion takes place in a water-containing environment which avoids standard drying processes after harvesting. The ability of HTL to deal with high levels of water is considered as most relevant (Milledge et al. 2014), though some problems could arise for values higher than 90 wt%, due to unfavourable energy balance (Vardon et al. 2012).

3.2.2 *Effect of process conditions*

In general, it is recognised that temperature presents a significant effect on conversion rates of algae biomass and composition. When depolymerisation occurs, concentration of free radicals tends to increase, leading to repolymerisation of fragmented species. During fragmentation and repolymerisation, temperature effect will be critical to final outcome, as depolymerisation reactions are dominant in the initial stages of pyrolysis, while repolymerisation reactions tend to increase at later stages, as higher reaction times tend to increase the formation of char.

The suitable temperature for HTL conversion depends also on other process conditions (e.g., algae species and reactors types, among others). Moreover, the selection of optimum temperature yielding oils for a specific purpose may be a critical task, as water properties also change significantly near supercritical conditions. Even so, the rise of temperature tends to increase the formation of oil-compounds, though its continuous increase will inhibit biomass liquefaction. Two main reasons could explain this behaviour: secondary decompositions become more active at high temperatures, which tends to increase the formation of gaseous compounds, and recombination of free radical reactions leads to char formation, due to their high concentrations. As these mechanisms become dominant at high temperatures, the conversion of algae into liquid oils will not be favourable. Lower temperatures may also result in incomplete conversion even when lignin and cellulose fragments are rapidly decomposed at temperatures of 250°C (Akhtar and Amin 2011).

Pressure is another parameter that may influence the conversion rates during hydrothermal liquefaction. This parameter maintains single-phase media, preventing large enthalpy inputs that are required for phase change. Pressure is dependent on the temperature chosen. By keeping pressure at suitable values, favourable reaction pathways of hydrolysis and biomass dissolution may be controlled towards the formation of liquid and gaseous compounds. Moreover, pressure also increases solvent density while high-density medium favours the conversion.

The effect of reaction time as well as catalyst load (type and amount) are also important when considering algae feedstock as a reliable technology for biofuel production. The effect of catalyst presence was already addressed. Some authors suggested that in continuous-flow systems, very short residence times produces bio-oil yields similar to longer residence times in batch reactors (Douglas et al. 2015). In addition, more severe reaction conditions tend to increase bio-oil yields with lower oxygen content, although with higher nitrogen content.

The objective of particle size reduction is to achieve a higher degree of hydrolysis and fragmentation of biomass. However, in hydrothermal liquefaction of biomass, particle size is normally considered as a less important parameter, as no significant impact on the yield of bio-oil has been reported. These results may be due to the presence of sub/supercritical water, which act as a heat transfer medium and also as an extracting medium, which makes particle size a minor parameter. So, the heat transfer limitations in hydrothermal liquefaction are overcome by the presence of sub/supercritical water (Akhtar and Amin 2011). Also, the size reduction of biomass presents high energy costs. Thus, to compensate this increase of costs, the optimum particle size of biomass must increase the yield of products of hydrothermal liquefaction with low grinding cost. Zhang et al.

(2009) studied the effect of three different particles sizes (25.4 mm, 2 mm, and 0.5 mm) on the liquid oil yield in hydrothermal liquefaction of grass. They did not observe any improve in liquid oil yield at 350°C with the reduction of the particle size. So, frequently is recommendable the use of a particle size distributions with the lower grinding costs. According to Mani et al. 2004, the grinding cost almost doubled for particle size reduction from 3.2 mm to 0.8 mm during dry grinding of wheat straw, barley straw, and switch grass. Hence, Jindal and Jha concluded that the particle size, which overcomes heat and mass transfer limitations in hydrothermal liquefaction, at reasonable grinding cost, should be between 4 and 10 mm (Jindal and Jha 2015).

A significant number of studies found in literature suggested that process conditions could range between 280 and 375°C, pressure values from 4 to 22 MPa and reaction time from 5 to 120 min should be used. Shuping et al. 2010 reported 360°C as the optimum temperature for microalgae *Dunaliella tertiolecta,* while Zhoufan et al. 2015 reported higher liquid yields at 310°C. Barreiro et al. 2013 reviewed a number of studies based on microalgae HTL for maximum oil yield and reported that liquid yields could range between 20 to 64% depending on the process conditions.

3.2.3 *Products and perspectives*

Algae biomass has received high level of interest as a renewable resource for biofuel production, mostly in regard to its direct pathway for liquid bio-oil production through HTL and available amounts due to relatively high growth rates. The main HTL product, which is a viscous liquid at room temperature, is a complex mixture of aromatic and oxygenated hydrocarbons, as well as heterocyclic and nitrogenated compounds (Elliott et al. 2013). When compared with some lignocellulosic biomasses products, algae bio oils usually present lower density values, less dissolved water and lower acid content, while nitrogen and sulphur amounts may be found in higher amounts. Although this biofuel resembles to fossil petroleum with respect to energy value (Jena and Das 2011), physical and chemical characteristics are not suitable for its direct use as transportation fuel. Those limitations include undesired high N contents found in bio oils, which could result in NO_x emissions during direct combustion, as well as the presence of cyclic nitrogen compounds which tend to be problematic for some bio oils utilisations. Due to its physical and chemical nature, algae HTL bio oils require additional treatments (e.g., processing and upgrading) to meet current fuel-use requirements. Those could include solid particles removal by settling and filtration and different processes to improve bio oils properties to different grades of fuels, namely: distillation, catalytic hydro-treating (reduction of O, N and S levels), catalytic cracking, decarboxylation, aromatization, and blending (to reduce viscosity). Due to the early stage of development, it is expected that new research and technological developments may provide solutions for bio oils direct inclusion in existing petroleum refining infrastructures.

As bio-oil production only includes a small fraction that reaches the properties of standard fuels, one possible solution may result from blending them with petroleum derived products, prior to its fractional distillation in the conventional refinery (although some current specifications are still required to meet in respect to N and S levels). The economic viability of the process is a critical issue to its future implementation. Based on available information, a number of simulation models and cost analysis have been developed. Those models include products yields, raw materials consumption, carbon efficiency, and energy efficiency.

According to publications, a number of researchers have attempted to correlate bio-oil yields to algae composition, and the findings suggested that overall yields could not be based on a simple average yields obtained from model compounds. From existing research, it seems that reaction of protein and carbohydrates do not contribute individually for the bio-oil formation. When reactions occur in complex mediums, interaction effects could strongly influence final products. In fact, algae biochemical structure seems to dictate the bio-oil quality and yields while, direct correlations do not fully explain the findings. Although reasonable correlations may be found in respect to lipid content and bio-oil yields, much weaker correlations are found for protein and carbohydrate contents. In general, reported bio oils yields were higher than those expected. This situation could be attributed to proteins and carbohydrates decomposition, leading to amino acids and sugars, which could further react to form nitrogen containing cyclic organic compounds (pyridines and pyrroles). Although the lipid based models are easier to implement, the presence of carbohydrates and protein gives rise to the Maillard-type reactions, which could result in multiple pathways for product formation. Different levels of salts and metals found in algae could also affect the type and yields of products obtained. Even so, different reactions such as decarboxylation, denitrogenation, repolymerisation, and condensation may occur simultaneously to decompose, rearrange, and form new molecules during different stages of the HTL reactions. Although these reactions are complex and not fully understood under HTL conditions, some authors suggested that nitrogen-containing heterocycles are formed through these types of reactions (Wagner et al. 2016). A number of attempts have been made to predict the final outcome based on the algae biochemical composition (Shijie et al. 2015).

Primary focus on research has been the recovery of the fatty acid triglycerides obtained from algae HTL as a feedstock source for biodiesel production (although not all are high fatty acid producers). Even considering those species that present suitable characteristics, additional thoughts regarding grown controlled conditions are required when fuel production is of concern. Some strategies to overcome this problem may include algae grow in a wild and/or mixed culture in order to maximize total biomass without considering the maximization of only fatty acid production. Through an adequate balance in conversion of most suitable biomasses (based on carbohydrates, protein and lipid structures), bio-oil production from different sources may be a sustainable option in the near future.

Nutrient availability is one of the critical issues for algae growth. According to algae type and growth conditions, some differences may be found in carbon (C), nitrogen (N), and phosphorus (P) typical ratios. To date, much of the research has been focused on N contents and, to a lesser extent on P, and its relation with C based on algae growth location or lipid accumulation. Different results were obtained, depending on the algae type. The supply of C (either as bicarbonate (HCO_3^-) or CO_2) is critical in the growth cycle, including the increase of both algae productivity and lipid content. Other nutrients such as silicon (Si) and iron (Fe) still require further research in respect to C/N/P ratios and allocation of C into the formation of lipids. According to some researches, bio-oil yields obtained from algae HTL process tend to be higher than those found from original oil content of the algae, suggesting that the conversion to liquids not only occurs from lipids molecules but also from the non-lipid components of the algae biomass. Recent research studies suggested that lipids act as a reservoir for polyunsaturated fatty acids (PUFAs), which play a key role in the structural components of cell membranes. In addition, lipids are also considered as a potential energy-rich biodiesel precursors through transesterification process to produce fatty acid methyl esters (FAME), which is a biological equivalent to diesel fuel (Fields et al. 2014). However, it was also reported that PUFAs content in lipids could negatively impact biodiesel quality, and some findings also suggested that not only lipids amounts, but lipid composition should also be considered (Shekh et al. 2013).

Although water and light are also critical to algae grow, to date, different groups of researchers have reported deep limitations to its economical use for energy production, mostly in regard to high costs of harvesting and concentration, as well as product sustainability (Ozkan et al. 2012, Schnurr et al. 2013, Bernstein et al. 2014). The integration of life cycle analysis (LCA) to algae biofuels production could highlight major pathways that produce a net increase in system performance. Ongoing application of LCA may also lead to advances in research that promote the development of more cost-effective and environmentally-friendly biofuel production processes. Despite greater capital expense of HTL, absence of drying or dewatering requirements, as HTL allows the use of wet feedstock, and total algae conversion (yields of around 60%) into useable bio oils are currently considered as main advantages (Liu et al. 2013).

When attempting to assess the economic competitiveness of algae into biofuels production at industrial scale, different sources of uncertainty have been reported in literature, which are grounded on different lab- and pilot-scale data. Given the different assumptions, an in-depth analysis between multiple studies is still required, based on a more direct comparison of similar criteria and objectives.

In Fig. 5.7, a simplified conceptual scheme for algae HTL is presented. Usually, process schemes are drawn mostly on simple basis, while completely defined process description (including energy and mass balances calculations) are still lacking or in a very early stage of development. Although the production of algae into bio-oil has been developed in some pilot-scale facilities, very limited information could be found in respect to whether algae fuels can be effectively produced in sufficient amount to displace petroleum fuels. Limitations to commercialization need to be well understood and addressed for future implementation. When compared to petroleum derived fuels, algae derived bio oils still have some drawbacks, as they are still very expensive and improved economics of production in a large-scale framework are not sufficient for an environmentally sustainable

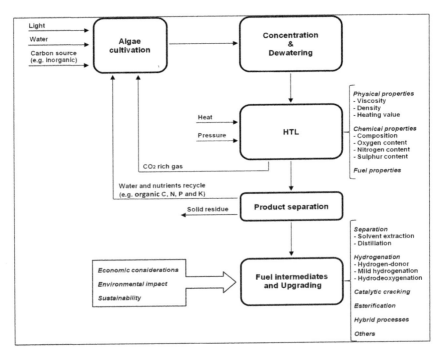

Figure 5.7. Conceptual scheme of algae hydrothermal liquefaction (HTL).

and feasibility production. Bio-oil chemical composition may consist of a wide range of organic compounds (acids, alcohols, aldehydes, esters, ketones, phenols, and lignin-derived oligomers) although some of them may be undesirable for fuel purposes. Thus, upgrading post-treatment is required by common physical and chemical and/or thermochemical processes. Prior to post-treatment processes, the production of some contaminants may be reduced by using catalyst suitable for selective degradation. Moreover, the use of selective catalyst is currently crucial to enhancement conversion rates of algae, to shorten reaction time and consequently, to reduce energy requirements and production costs. Most common catalyst used for these purposes includes both homogeneous and heterogeneous catalysts, as mentioned before.

Depending on algae source and HTL conditions, bio-oil relative composition could include triglycerides, terpenoid hydrocarbons, polar lipids, oxygen-containing nontriglyceride carotenoid oils, and chlorophylls. All of these compounds are known to be both energy and carbon rich, which contribute for the high energy content of bio oils (around 35.8 $MJ.kg^{-1}$). Algae bio-oil seems to be rich in long chain polyunsaturated fatty acids, which may not be as suitable for biodiesel production. Nevertheless, some findings suggested that algae bio-oil could be successfully transformed into different combustible products which are very similar to equivalent petroleum derived fuels (e.g., kerosene, gasoline, and diesel). Moreover, different groups of researchers, on the topic of production of liquid fuels from algae, strengthen the view that algae transport fuels are a proven technology, although unaffordable at the present (Yusuf 2013).

As stated before, a number of studies have been conducted on HTL algae and some progresses have been successfully accomplished towards biofuel production. Nevertheless, new research works and developments on the topic are still required for fuel purposes in regard to: storage and stability of biofuel, large-scale reactor design and best operation conditions, algae feedstock and blending conditions, and algae type and growth conditions.

3.2.4 *Equipment and constrains*

Although different algae sources are available with significant potential for bio-oil production, from the engineering perspective other issues are also of concern in which are included high processing energy requirements. To date, general concerns are based on a lesser extent on the definition of best solution and operation conditions, but mostly on the current low economic value of the bio-oil in comparison to some high value products obtained from algal biomass. In respect to the process conditions, high demand of heat is required that needs to be further recovered by heat exchangers. In addition, algae composition will be an important issue to consider due to the high protein content, which could result in high N levels. Unconverted material and/or solid residue should also be considered.

Bio-oil physical characteristics may result in a paste-like consistency flow that behaves as a non-Newtonian fluid. This high viscosity fluid will lead to poor heat transfer conditions, which could significantly change the initial operation conditions requirements. Yields and bio-oil physical and chemical properties may be affected, while char and gaseous compounds tend to increase as process continuous during longer residence times and carbonisation is favoured. Though some problems associated to carbonization effects and longer residence times may be minimized through larger reactors, both higher costs and higher thermal energy losses by convection and radiation are more difficult to address properly. Reactor configuration will also account for the residence time, in which algae will be subjected to conversion. Significant differences may also be found in literature, ranging from 30 seconds up to 40 minutes, making it more difficult to determine the best configuration for suitable-size large scale HTL reactors (Elliott et al. 2015). Thus, additional research is still required on different topics such as reactor design and kinetics, process energy consumption, environmental impact, bio-oil yields, product treatment and upgrading, as well as economic evaluation in order to provide more reasonable base-information for process design and scale-up for commercialization.

Most research on HTL and its application to algae bio-oil production has been primarily related to batch reactor tests (Chow et al. 2013). Currently, semi-continuous and continuous-type reactors may also be found. In order to increase the economic competitiveness of algae HTL into fuel markets, additional technological developments are required on equipment and materials (e.g., resistant to corrosion). Although the current technology is still in its early stage of development, the economic viability will not be achieved through batch-type reactors, as they are probably inadequate for a low-cost biofuel industrial-scale production. Continuous-type reactors allow high productivity with less energy consumption when compared to batch-type reactors. In particular, fully operational reactors should be able to operate during long-term under very demanding operating conditions. Glass reactors used in laboratory tests are resistant to chemical corrosion, but some disadvantages may be found in respect to performance at high temperatures and pressures. As a result, the utilization of glass reactors in algae HTL process will not be a valid option. Metal reactors present good resistance to high temperatures and pressures, although they may not be resistant enough to long chemical action periods unless more resistant alloys are used. For instance, stainless steel industrial equipment present low resistance to continuous corrosion under subcritical water conditions (Watanabe et al. 2004). The density of water and the density-related physical properties of water are well known as being fundamental issues to consider in regard to metal corrosion in near-critical or supercritical conditions.

Prior to bio-oil production through HTL process, additional constrains could also be considered in regard to large-scale culture technologies for producing algae biomass. Those include: most suitable algae species and constrains in regard to productivity, location and design of optimal cultivation system (pond geometry), surface-to-volume ratio to maximize sunlight capture and its transmission to algae, availability of CO_2 (each ton of algal biomass requires 1.83 tons of CO_2), adequate supply of water and nutrients (e.g., organic C, N, and P), and energy consumption of pumping and recovery systems (e.g., flocculation-sedimentation, centrifugation, and filtration). Until now, the most common used are raceway ponds, which present low productivity (0.5 to 1.5 $kg.m^{-3}$) when compared to photobioreactors (5 $kg.m^{-3}$) being the last one more expensive and high energy demand to operate. In respect to biomass to bio-oil ratio, existing culture systems do not come close to the biological limits of productivity as algae culture is light limited. Physical limitations could be found in regard to mutual cell shading and photoinhibition at light levels lower than 10% of peak midday sunlight level (tropical regions). Theoretical models estimate biomass productivity as being around 0.095 $kg.m^{-2}.d^{-1}$ (for 40 wt.% oil in biomass), although culture limitations, such as raceways, may reduce biomass productivity to lower values (0.025 $kg.m^{-2}.d^{-1}$). Biomass productivity will be even lower if algae oil content is higher (Yusuf 2013). Whether raceways can be used to produce sufficient algal biomass to be converted into bio-oil at economically viable large-scale facilities are still debatable.

3.2.5 Critical considerations to commercialization

The need of alternative fuels is mostly driven by energy demand, energy security, CO_2 emissions, and global warming considerations, due to the consumption of fossil fuels. Currently, fossil fuels are cheaper and readily available than majority of other solutions, nevertheless finding new solutions and/or develop new technologies has become critical to minimize environmental impact and to diversify the supply of fuel sources.

Although a significant number of research works have been published in regard to algae HTL, there is still limited knowledge on chemistry that takes place on this complex reactional medium. More research on the topic is still required, including kinetics and catalyst materials in order to understand the pathways of chemical reactions and compounds formation, to target specific products and minimize the formation of undesirable compounds. Moreover, when considering full-scale commercialization facilities additional engineering challenges are still to overcome mostly in respect to large-size reactors and materials that withstand high temperature and high-pressure conditions in corrosive environments.

Currently, there are a number of critical issues hindering commercialization that require solving for ultimately scaled up implementation. Those include solids loading—in a number of situations significant findings are reported involving higher conversion rates and high quality products at low feed concentrations; if those findings are not observed during scale-up to higher capacities, they are of little commercial value; feedstock impurities—most feedstock present several impurities, even if they have low contents; truly low-cost feedstocks are usually mixed with a number of impurities and treatment techniques could become crucial and expensive; heat transfer and recovery systems—HTL conditions operate at high temperatures and high heating rates; the integration of heat recovery systems with high efficiencies is of concern; algae feeding—feeding a mixture of algae and water into a reactor operating at 22 MPa is engineering challenging; thus, high-pressure pumps with high flow rate capacities are required; presence of catalyst—although homogeneous catalysts do not pose coking or inactivation problems, with heterogeneous catalysts fouling effects may be found, which increase catalyst deactivation; a great number of studies with demonstrated high yields of desired products in batch or short-time continuous reactions are reported, although long-time periods often result in significant reduction in catalyst effectiveness (e.g., precipitation of inorganics, degradation of catalyst support, and oxidization); wall effects—the majority of research has been conducted within Inconel and Hastelloy reactors, normally composed by high content of nickel; and the nickel present in wall surfaces could act as a catalyst, making it difficult to evaluate the real effect of the chosen catalyst. This issue has been reported by different groups of researches, and is considered to be critical when commercialization is of concern (Peterson et al. 2008).

In addition to the above mentioned points, the major difficulty is actually producing a large quantity of algal biomass in both sustainable and economical ways, based on current petroleum refining prices. Although price of petroleum fuels have reached high values and resembles little the current producing costs, price fluctuations pose significant limitations to investment in a large-scale industrial facility of algal biomass as source of biofuel. Even so, a significant reduction in carbon footprint is attributed to algae (when compared with petroleum fuels), which could be a strong driver for its implementation. The price at which algae biofuel can be produced will be a major issue to the stakeholders' commitment in the commercialization.

3.3 Hydrothermal Carbonization

The main product of HTC is biochar, as shown in Fig. 5.1, which is similar to charcoal, though produced from biomass. HTC main products are solid, which are may be referred as biochar or hydrochar, due to the presence of water. The production of charcoal is several million tons per year. Research findings during the past decade have developed a renewed interest in this product. This development was due, especially to the development of pressurized equipment that improved the yields of charcoal and reduced significantly the reaction times needed (Antal and Gronli 2003). However, due to the "energy crisis", the research on conventional thermochemical processes (pyrolysis and liquefaction) using biomass were, mainly, focused in the production of pyrolytic liquid fuels (i.e., biocrude oil) minimizing the gases and char product, that were considered by-products. These processes were, mostly, applied to dry feedstocks. When the feedstock used has high water content (i.e., moisture higher than 60%), hydrothermal processes are more appropriate.

One of these processes is the hydrothermal carbonization (HTC), which is a relatively new method and has gained importance in recent years (Tekin et al. 2014). This process allows the utilization of several alternative non-traditional feedstocks: municipal wastes (MSW), wet animal manures, human waste, sewage sludge, wet agricultural residues, and also aquaculture and algal residues. The main advantage of HTC is that it can convert wet feedstock into carbonaceous solids at quite high yields without needing energy-intensive drying of feedstock before or during the process (Libra et al. 2011b).

During this process, the organic matter in the biomass is thermochemically decomposed by heating in the absence of oxygen and at subcritical liquid water conditions. The relative proportions of the gas, liquid, and solid phase, their characteristics and process energy requirements depend on the input material and on the process conditions. In hydrothermal carbonization, temperature in the range of 150–350°C and pressures from 4 to 5 MPa are typically used (Basso et al. 2016, Jain et al. 2016). The process temperature is dependent on the type of raw materials used and its decomposition temperature.

The most influent operational conditions in the process and the ranges of value that need to be defined are Funke and Ziegler (2010):

- Operation should be limited to subcritical conditions of water, due to physical and chemical reasons.
- It is necessary that the presence of water is in liquid phase, so at least saturated pressure is needed. During the HTC process, water acts as a solvent and as a catalyst facilitating the hydrolysis and cleavage of the lignocellulosic biomass.
- It is imperative that the feedstock is submerged in the water during the whole process.

- The temperature must be above 100°C, once at this range the first reactions will take place. The significant hydrolysis reactions start at temperature of about 180°C.
- Once alkaline conditions lead to a substantially different product, the pH value should be below 7. However the pH value drops automatically during hydrothermal carbonization process, due to the formation of by-products.
- Reaction times normally used varied between 1 and 72 h.

3.3.1 Main hydrothermal carbonization reactions

During HTC are produced solid, gaseous and water soluble organic compounds. The solid compounds are produced in higher concentrations while the gaseous and water soluble organic compounds are present in lower amounts. The main gaseous compound formed is CO_2.

Biomass contains different polymer chains, so the mechanism for the conversion to hydrochar is rather complex. Hemicellulose, cellulose, and lignin are considered to be the main constituents of typical biomass. The hydrolysis of hemicellulose content in biomass occur at lower temperatures (about 200°C) and lead to the formation of hydrochar (Jain et al. 2016, Kang et al. 2012). Regarding lignin decomposition, it was reported that the conversion to hydrochar occurs at a higher temperature (Falco et al. 2011). Kang et al. 2012 stated that lignin slows the release of decomposition products formed from polysaccharides, interfering with the hydrolysis of cellulose and hemicellulose. So, lignin seems to contribute to maintaining macrostructure of the initial biomass into the hydrothermal carbon products.

In literature, the most mentioned chemical reactions that may take place during hydrothermal carbonization, is the hydrolysis of cellulose (Funke and Ziegler 2010, Sevilla and Fuertes 2009b, Jain et al. 2016). For instance, Sevilla and Fuertes 2009b proposed the formation of char from carbohydrates/cellulose through polymerization or condensation of dissolved monomers produced from the biomass decomposition. These mechanisms include hydrolysis, dehydration, decarboxylation, condensation polymerization, and aromatization reactions.

The hydrolysis reaction leads to the cleavage of chemical bonds of mainly esters and ethers of the macromolecules by the addition of 1 mole of water. The cellulose hydrolysis takes place, mainly, under hydrothermal conditions above approximately 200°C. The dehydration reaction can include chemical reactions and physical processes. The physical processes remove water from the biomass matrix without changing its chemical constitution. The chemical dehydration carbonizes significantly the biomass by reducing the O/C and H/C ratios. Cellulose begins to decompose by pure dehydration according to the reaction (27).

$$4 \, (C_6H_{10}O_5)_n \Leftrightarrow 2 \, (C_{12}H_{10}O_5)_n + 10 \, H_2O \tag{27}$$

The substantial decarboxylation only takes place after a specific amount of water being released. The carboxyl and carbonyl groups degrade rapidly above 150°C, forming CO and CO_2 by reaction (28). However, the results obtained by Funke and Ziegler 2010 showed that more CO_2 was formed than can be expected by the elimination of carboxyl groups. Thus other mechanisms should be involved.

$$\overset{H_2O}{\underset{H^+ \text{or } HO^-, \, heat}{RCOCH_2COOR' \Leftrightarrow RCOCH_3 + CO_2}} \tag{28}$$

One of the possible sources of CO_2 is formic acid, which is formed in substantial amounts during the degradation of cellulose and decomposes under hydrothermal conditions, forming mainly CO_2 and H_2O. Also, CO_2 can be produced during the condensation reactions and by the cleavage of intramolecular bonds. Other supposition is that H_2O acts as an oxidizing agent at high temperatures (above 300°C), and so additional CO_2 is formed by the thermal decomposition of such oxidized elements.

The polymerization reactions are due to presence of unsaturated compounds, which polymerize easily, that were formed by the elimination of carboxyl and hydroxyl groups. This fragments formed from degradation of macromolecules, under hydrothermal conditions, are highly reactive. The condensation reactions, for instance, of the aromatic ring are frequently responsible for the formation of some amount of CO_2. Also, the possible condensation and cleavage reactions of relevant molecular groups can explain the experimental results obtained during HTC in relation to the influence of reaction temperature and pressure on product yields and composition.

Moreover, the rate of carbonization is increasingly influenced by the higher condensation degree of aromatic compounds. Therefore, it may be concluded that the formation of solids during HTC, by hydrothermal carbonization, is mainly due to the condensation polymerization reactions, specifically aldol condensation (Funke and Ziegler 2010, Jain et al. 2016). At subcritical conditions, the free radicals are successfully saturated by the water present and hydrogen donated by the aromatization reactions. In the literature, detailed information was not found about polymerization sequences during hydrothermal carbonization, once the solid formation by polymerization is normally considered an unwanted side reaction and so, is not referred in other hydrothermal processes.

Regarding aromatization reactions, the (hemi-)cellulose consists of carbohydrates, but it is able to form aromatic structures in hydrothermal conditions and the aromaticity of carbon structures by hydrothermal carbonization rise with the increase of reaction severity. The aromatic structures show high stability under hydrothermal conditions and so, they can be considered as the basic unit for the formation of the HTC solids. So, it is highly possible that the effect of hydrothermal treatment on the carbon content is reduced with the increasing of aromatic bonds.

Other reactions are possible to occur, for instance, transformation reactions within the lignin molecule instead of fragmentation followed by polymerization. However, in these reactions, the physical structure of the biomass is kept considerably long. Given the high rate of fragmentation observed during the degradation of the biomass structure, due to hydrolysis reaction above 180°C, it seems improbable that these reactions play an important role in the process.

The pyrolytic reactions have also been reported as competing reactions under hydrothermal conditions. However, these reactions might only occur above 200°C, and the pyrolysis products have not been detected in significant amounts during HTC. Nevertheless, the pyrolytic reactions have been used to explain the carbonization of the macromolecules that were not in contact with water, because they were blocked by the precipitation of condensing fragments.

The Fischer-Tropsch type reactions have also been detected under hydrothermal conditions and can have an important role, due to the amount of CO_2 being produced in HTC. However, so far, the role of these reactions has not been studied in detail.

3.3.2 Hydrothermal carbonization products

A vast number of intermediate products involved in the hydrothermal carbonization were identified, which are the result of the multiplicity of the different mechanisms involved. However, there are three main products that are important to refer, solids, liquids, and gases, whose distribution heavily depends on process conditions and feedstock used.

The solid fraction, the main objective of HTC, has aroused an increasing interest in the latest research. This solid product, often called hydrochar, is similar to natural coal, and it is characterized by an agglomeration of different chemical substances. The basic characterization, elementary analysis of this product, shows that its composition may approach that of lignite or even sub-bituminous coal, depending on reaction severity conditions (Funke and Ziegler 2010).

The liquids products are mainly composed of water and a high percentage of inorganic and organic compounds (sugar and derivatives, organic acids, furanoid, and phenoic compounds), where many of them can be potentially used as valuable chemicals. However, the liquid fraction will always contain relatively important amounts of solid particles which cannot be recovered easily. These solid particles are dispersed solid particles in the liquid. The formation of a second immiscible liquid phase during hydrothermal carbonization has been reported; even when higher temperatures are used, this liquid is similar to the oil generated in hydrous pyrolysis.

The gaseous fraction is mainly composed by CO_2, though small amounts of CO, CH_4, and H_2 as well as traces of C_mH_n were also detected (Basso et al. 2016). It is also likely that the same amount of CO_2 present in the gaseous phase is dissolved in water. The studies performed refer that the amount of gaseous products increases with the increase of the reaction temperature and at the same time, at these conditions the fraction of CO decreases, while the content of CH_4 and H_2 increases.

3.3.3 Influence of experimental parameters on HTC process

The properties of hydrochar, especially the oxygenated functional groups, are dependent on the raw material used, and the degree of reactions is dependent on the combination of temperature, reaction time, substrate concentration, type of catalyst used, and pH. These process parameters can influence the properties of hydrochar, as they affect the development of high quality activated carbons. The oxygenated functional groups content in hydrochar is an important factor that can influence the degree of porosity in the activated carbon, so it is a parameter evaluated in the studies of HTC.

Several studies on HTC have been carried out to study the effects of different experimental conditions on the product yield and quality (hydrochar, liquid, and gas) (Titirici et al. 2012, Berge et al. 2011, Sevilla and Fuertes 2009a, Romero-Anaya et al. 2014, Hoekman et al. 2011, Libra et al. 2011a, Sun et al. 2014).

Reaction temperature

In hydrothermal carbonization a temperature in the range of 150–350°C is typically used. The reaction temperature, in hydrothermal carbonization process, is an important and controlling factor in determining the type of degradation reactions that take place, and consequently the yields and quality of the products formed. For instance, when high temperatures are used, a wider range of products and higher gas yields are formed. Also, the relative carbon content increases and the relative oxygen concentration usually decreases with increasing reaction time and temperature. Typically, at higher temperatures, the formation of more gaseous products is observed, and at lower temperatures more solid products are produced. With an increase in process temperatures up to 220°C with a pressure reaction around 20 bar, approximately 1–5% of gas is typically produced, and most of the organic material in biomass is converted into solids (Jain et al. 2016). Higher temperature also leads to wider dehydration and to an increase in the degree of condensation of the hydrochar. These results were observed by Sevilla and Fuertes 2009a, which detected a decrease in the O/C and H/C atomic ratios with the increase of the temperature of the hydrothermal carbonization from 230 to 250°C. In another study, Kang et al. 2012 stated a decrease in ion exchange capacity with the rise of temperature (from 225 to 265°C) for cellulose, lignin, and lignocellulosic biomass, and observed a reduction in the oxygenated functional groups and in the hydroxyl groups with the increase of temperature for all the starting materials except lignin (Kang et al. 2012). In the case of lignin, Dinjus et al. 2011 observed a small increase in the oxygenated functional groups when the temperature increased from 225 to 245°C, due to high temperature conditions for the hydrolysis of lignin.

Jain et al. 2016 also investigated the effect of hydrothermal carbonization temperature (200, 275, 315, and 350°C) on the oxygenated functional groups formation using coconut shell based hydrochars. The maximum value was observed at 275°C. At 315 and 350°C, the oxygenated functional groups content decrease, probably due to the higher extent of decomposition of these functional groups and to the formation of gaseous products at higher temperatures (Marsman et al. 2007, Williams and Onwudili 2005). On the other hand, the oxygenated functional groups are also dependent on the type of feedstock, residence time, and substrate concentrations used. For instance, when high temperatures are used in the lignocellulosic biomass treatment, the high yields of oxygenated functional groups can be achieved at relatively low reaction times, perhaps due to higher reaction rate, and therefore easy hydrolysis (Dinjus et al. 2011). Also, high yields of oxygenated functional groups can be obtained when higher substrate concentrations are used, because hydrolysis becomes easier when higher amounts of biomass per gram of water are used. Regarding the feedstock, higher lignin content also increase the oxygenated functional groups content, due to higher reaction severity, which make easier the disintegration of the biomass into smaller constituents (Cao et al. 2013). These results seem to indicate that knowing the initial composition of the feedstock, the hydrothermal conditions can be adjusted in order to obtain the hydrothermal carbonization products for various applications.

Temperature can also influence, indirectly, the hydrothermal carbonization, by changing the properties of water. The water solvent properties are significantly improved at high temperatures. Moreover, the viscosity of the liquid water changes, depending on temperature between

0°C and 350°C, which allows an easier penetration of porous media with the improvement of the biomass decomposition.

Reaction time

The effect of the reaction time is closely connected with the reaction temperature and with the type of feedstock. For instance, it can be noticed that cellulose carbonizes at higher temperatures or at a lower temperature (for instance 200°C) after a longer reaction time. However, the effect of biomass feedstock is of utmost importance, and thus different results may be obtained with different materials.

Residence time during hydrothermal carbonization has an important role in the extent of reaction and in the quality and distribution of different types of products (Jain et al. 2016). Hydrothermal carbonization of biomass is described has a slow reaction. In the literature, it was found that residence time may range from a few minutes to several days. A longer residence time generally increases reaction severity, leading to more condensed products, due to excessive polymerization, which can result in a decrease in oxygenated functional groups. Several authors have studied the effect of reaction time on the oxygenated functional groups content (He et al. 2013, Zhang et al. 2014, Cao et al. 2013). He et al. 2013 reported a decrease in the oxygenated functional groups content when reaction time increased from 4 h to 12 h. However, a peak was observed at 6 h, followed by a decrease maybe due to excessive carbonization or its decomposition to gaseous products at higher reaction times. Nevertheless, the effect of reaction time in oxygenated functional groups is not linear, because the results reported in literature seem to indicate a complex dependency of oxygenated functional groups on temperature, reaction time, concentration and type of feedstock (Jain et al. 2016). Regarding the O/C and H/C atomic ratios, at longer reaction times high condensed products are expected, due to a higher extend of hydrolysis and polymerization reactions (Sevilla and Fuertes 2009b). Cao et al. 2013 also observed lower H/C ratios in sugar beet hydrochars and a slight decrease in O/C ratios in bark hydrochars with the rise of residence time from 3 to 20 h (Cao et al. 2013).

So, longer residence times lead to a decrease in hydrochar yields (grams of primary and secondary hydrochar/grams of biomass) (Hoekman et al. 2011, He et al. 2013). On the other hand, the increase of the reaction time resulted in higher formation of secondary char along with polyaromatic char from non-dissolved lignin content (Hoekman et al. 2011, Funke and Ziegler 2010). For instance, Hoekman et al. 2011 noticed a small decrease in the mass recovery with the increase in the residence time at 255°C (Hoekman et al. 2011). These results may be explained by the combination of the effect of the increase of polymerization and of the formation of secondary hydrochar, together with the decomposition of biomass to liquid/gaseous products. Thus, the biomass decomposition and condensation polymerization seems to rule the overall rate of the hydrothermal carbonization reactions (Jain et al. 2016).

Pressure

The influence of pressure on HTC reactions is relatively minor compared to that of temperature, though the pressure attained in the reaction medium is highly dependent on temperature used (Jain et al. 2016). The reaction pressure influences the reaction scheme according to the principles of Le Chatelier. When the reaction pressure rises, the reaction equilibrium shifts to solid and liquid phases, and both dehydration and decarboxylation as primary overall reactions lose importance. However, it was observed experimentally that pressure, isolated from temperature, has low impact in the HTC reactions (Funke and Ziegler 2010).

Substrate concentration

It is known that biomass above the liquid surface does not carbonise; this fact is also mentioned in the studies on hydrous pyrolysis experiments (Funke and Ziegler 2010). However, it is possible to carbonize biomass in oil or in water, but carbonization process is faster when water is used (Funke and Ziegler 2010). Regarding the concentration of the reactants, the reduced concentrations in the hydrothermal environment reduce the cross-reactions of the involved species, and leads to a more well-defined product. At higher concentrations, it has been observed only slight changes in the production of liquid products and the polymerization becomes the main process (Funke and Ziegler 2010). Hence, the reactant-to-water ratio is an important parameter for the HTC process. When this ration is higher the polymerization takes place at quite shorter residence times, but at the same time the percentage of reactant that is not hydrolysed is higher. This result was observed by Sevilla and Fuertes 2009b, who reported minus condensed products (high O/C and H/C atomic ratios) at higher substrate concentrations, due to incomplete hydrolysis (Sevilla and Fuertes 2009b). Therefore, the residence time and the substrate concentrations can be adjusted according to the type of product needed. For instance, higher concentration coupled with a shorter residence time will promote the production of liquid phase products. Regarding oxygenated functional groups, the high substrate concentrations leads to high oxygenated functional groups at relatively higher temperature, due to easier hydrolysis of higher amounts of biomass per gram of water at high reaction severity conditions. At higher residence times, the content of oxygenated functional groups also increases with the subtract concentration, due to longer exposure which leads to higher extent of hydrolysis. High substrate concentrations can also lead to high oxygenated functional groups at relatively lower lignin content, because higher concentrations biomass with higher lignin content will be difficult to disintegrate at a specific temperature and reaction time. Moreover, to achieve the maximum hydrochar yield, the use of high reactant concentrations is comprehensive as the complete conversion is enhanced at higher initial concentrations (Sevilla and Fuertes 2009b, Knezevic et al. 2009).

3.3.4 Microalgae HTC

The HTC has been mainly applied for the treatment of several types of wet biomass, but a few studies on the application of this process to microalgae were found. HTC at 250°C produces mainly solids and water, however for microalgae HTC the production of biocrude is also favoured, due to the significant level of oil contained in the microalgae. So, it was also observed a significant amount of oil in the hydrochar from the microalgae (around 20 wt.%). The gas yield was usually obtained in a range from 6% to 12% (Ekpo et al. 2016). The main objective of the study performed by Heilmann et al. 2010 was hydrochar production, in order to obtain a high level of carbonization and hydrochar yield by minimizing simultaneously the processing time. They stated that the microalgae should be excellent biomass substrates for hydrochar production, because their small size should facilitate the fast thermal transfer to processing temperatures. They reported the results obtained with green and blue-green microalgae, which are not lignocellulosic biomass. Thus, the chemistry involved is completely different, because it involves, mainly, proteins, lipids, and carbohydrates. However, Heilmann et al. 2010 stated that microalgae can be converted into algal hydrochar product that is similar to bituminous coal quality, using moderate conditions of reaction temperature and time (approximately 200°C and less than an hour) and pressure lower than 2 MPa. The same authors also studied a process for isolation of three products (fatty acids, chars, and nutrient-rich aqueous phases) from the hydrothermal carbonization of microalgae. Their main objective was to recovery the oil present in the hydrochar, and the process involves the HTC and a lipid extraction process. The results presented in this study indicate that fatty acids derived from hydrothermal carbonization of microalgae look very promising for the production of liquid biofuels (Heilmann et al. 2011).

4. Future Perspectives and R&D Needs

The valorisation of algal biomass through advance technologies such as HTC, HTL, and SCWG to produce biofuels presents some advantages, though research and development (R&D) is still needed. To date, most of research on this subject has been focused on the production of biofuels through conventional chemical and biochemical processes. With the ongoing laboratory experiments and small scale plants, a great number of studies have shown that algae biomass has great potential to be explored as a valuable source of feedstock towards the production of different biofuels and biochemicals. The use of advanced technologies that allows using the wet nature of algae will certainly be considered as a significant advantage, though additional research and engineering developments on the subject are still required. Those include the reduction of the environmental impact, the improvement of quality and of overall efficiency by selecting the best algae biomass (identification and processing), the integration of these technologies with current ones and the output of production plants.

Prior to production of biofuels and biochemicals, other major issues are also of concern when considering chemical, physical, and biological harvesting methods which may pose some restrains in choosing the most suitable algae feedstock. Those will result in differences mostly in capital and operating costs of algae harvesting and processing (e.g., energy consumption) as well as on land and location (geography), water, nutrients, and climate (e.g., temperature and light availability). Land is an essential commodity (whether bioreactors or ponds are used), and should not directly compete with agricultural needs. The location of growth and processing facilities are also crucial aspects to be considered. Water and nutrients are essential for any biological process, and the water recycle will be crucial as some locations are becoming increasingly stressed for potable water. Climate is also a critical factor, and is inherently related to temperature, as the need for light, energy, and different heat-regulation systems are required. Developing hybrid techniques, which make use of most suitable harvesting methods, may be a viable option that is worth exploring.

The large potential for converting algae biomass into a number of biochemicals for a variety of applications demands coordinated efforts between producers, end-users and regulators. Different target utilisation options are available, being most promising the ones related to biofuels and biochemicals production for different industries (e.g., petrochemical, chemical, and pharmaceutical). The production of biofuels and bio-based materials (including a wide variety of chemical precursors for specific purposes) will probably be the main issue to consider in existing petroleum refinery processes. These processes are capable of including new processes and/or adapting the current infrastructure, with some capital investment, to integrate advanced technologies. Moreover, the infrastructure for blending fuels as well as their testing and distribution structure is accessible in oil refineries. Some options are available in petroleum refineries, which are capable to include additional sources of feedstock. Those may include: distillation units, fluid catalytic cracking (FCC), and hydroprocessing units (hydrotreating and hydrocracking). In addition, the utilization of algae-derived synthesis gas (syngas) or hydrogen may also be an option to take into consideration.

Hydrothermal processes offer a unique way to obtain a wide range of biorefinery products. They can be considered as environmentally friendly processes, when compared with conventional petroleum-based fuels, using only water at different temperatures as a process medium. New pathways for integrated biorefineries to convert different algae-biomasses (feedstocks) into cost-competitive biofuels and biochemicals, in a flexible production scenario, will be a future goal. Moreover, the integration of technologies that enable the production of specific chemical building blocks and new materials to reduce the need for conventional fossil-based inputs present significant benefits. The viability and sustainability of algae as source of feedstock and its integration into common petroleum refinery processes for commercial purposes could rapidly decrease the current dependence on petroleum. Other major issue is the implementation by governments of financial contributions (e.g., subsidies or tax exemptions) to the production of all types of biofuels and biochemicals. With the expected continuous increase of price of petroleum-based fuels, the refining technology will be developed and/or adapted to allow the inclusion of different sources of algae biomasses in the production of bio- and petroleum-based fuels.

Besides the encouraging perspectives of using algal biomass to produce biofuels and biochemicals by wet or hydrothermal processes, there are still several barriers that need to be overcome.

- Price of algal biomass harvesting and/or growth
- Development of hydrothermal conversion processes to increase conversion rates, products yields, and quality
- Products upgrading research and development
- High integration of hydrothermal conversion processes with conventional petroleum refineries
- Process scale-up and fabrication
- Process demonstration

R&D for increasing conversion rates of hydrothermal processes and improving products yields and quality is essential to increase the technical viability of hydrothermal processes. Products upgrading and the development of multi-function catalyst that may increase process conversion and improve the selectivity towards desired compounds are also key issues.

Another barrier is process scale-up, as the good results obtained at lab and bench scale are sometimes difficult to replicate at higher scales. Afterwards, process demonstration should last for long periods of time to guarantee good and reliable operation to validate the process. It is also important to note that during the first stages of implementation of new processes like hydrothermal ones, it is important to have extra support like adequate incentives and policies, and even funding through governmental political measures and R&D strategies, supported by research financing programs to ensure the development of more efficient and cheaper processes.

The biorefinery concept has a very high priority in today's research and economics. Hydrothermal processing is very well-fitted for carrying out a breakdown of algae-biomass into a number of value-added products, starting from low temperature processes (HTC) to medium-higher temperature processes (HTL and SCWG). The combination of different hydrothermal processes, in sequences for producing specific products (tailored precursors with maximum yield and efficiency), and its integration in conventional petroleum refineries could be seen as a substantial advantage. This concept is based on a sequential processing chain of bio-based products and in the generation of higher-value products. It will possibly include large and/or small units (with savings in energy and transport costs). A realistic scenario in which conventional petroleum refineries cooperate with biorefineries could be based on the production of specific precursors for further processing and/or upgrading into current-used fuels, and on other petroleum-based products.

5. Conclusion

Hydrothermal processes seem to be quite adequate to deal with algal biomass, and especially to microalgae, as the very energy demanding processes of drying and concentration are not needed, once hydrothermal processes occur in presence of water, which behaves as a solvent, a reactant and a catalyst. Thus, while thermochemical dry processes need feedstocks with moisture contents below 20 wt%, hydrothermal processes may be applied to biomass with water content up to 70 wt% or even more. These processes may be applied to macroalgae and microalgae.

Algal biomass can grow in different types of environments not competing with food industry for water or arable land. They may be grown in integration with agricultural, animal farming, aquaculture and mineral processing systems, using the nutrients that still remain in the waters discarded by these processes, contributing to bioremediation of these wastewaters. Another option is the use of algae for purification of domestic wastewaters, having also an important role in carbon capture and sequestration (CCS). Algae may be used in different utilisations for instance in food, pharmaceuticals and cosmetics industries, but the huge amounts that are expected to be produced in the future, encourage the development of different end-uses, such as the production of biofuels and biochemical by hydrothermal processes.

Some authors have stated that microalgae could be used for hydrochar production by HTC, because their small size would facilitate the fast thermal transfer to processing temperatures. Nevertheless, other research studies have found that HTC is probably a best option, due to the high content of lipids found in microalgae that favour the production of bio oils.

There are still several barriers that need to be overcome before the use of algal biomass to produce biofuels and biochemicals by wet or hydrothermal processes. Most of them are related to improving processes conversion and increasing products yields and quality for which R&D is still needed.

Besides the encouraging technical results obtained so far, the concept of hydrothermal processes is proved at small scale. Thus, there is still a long way to run before the economic viability is fully proved. To achieve this goal, the construction of demonstration units is of most importance.

Keywords: Gasification; pyrolysis; hydrothermal carbonization; hydrothermal liquefaction; supercritical water gasification

References

Akhtar, J. and N.A.S. Amin. 2011. A review on process conditions for optimum bio-oil yield in hydrothermal liquefaction of biomass. Renewable and Sustainable Energy Reviews 15: 1615–1624.

Antal, M.J. and M. Gronli. 2003. The art, science, and technology of charcoal production. Ind. Eng. Chem. Res. 42: 1619–1640.

Araya, P.E., S.E. Droguett, H.J. Neuburg and R. Badilla-Ohlbaum. 1986. Catalytic wood liquefaction using a hydrogen donor solvent. Can. J. Chem. Eng. 64: 775–780.

Barreiro, D.L., W. Prins, F. Ronsse and W. Brilman. 2013. Hydrothermal liquefaction (HTL) of microalgae for biofuel production: State of the art review and future prospects. Biomass and Bioenergy 53: 113–127.

Basso, D., F. Patuzzi, D. Castello, M. Baratieri, E.C. Rada, E. Weiss-Hortala and L. Fiori. 2016. Agro-industrial waste to solid biofuel through hydrothermal carbonization. Waste Management 47: 114–121.

Berge, N.D., K.S. Ro, J. Mao, J.R.V. Flora, M.A. Chappell and S. Bae. 2011. Hydrothermal carbonization of municipal waste streams. Environ. Sci. Technol. 45: 5696–5703.

Bernstein, H.C., M. Kesaano, K. Moll, T. Smith, R. Gerlach, R.P. Carlson, C.D. Miller, B.M. Peyton, K.E. Cooksey, R.D. Gardner and R.C. Sims. 2014. Direct measurement and characterization of active photosynthesis zones inside wastewater remediating and biofuel producing microalgal biofilms. Bioresource Technology 156: 206–215.

Brennan, L. and P. Owende. 2010. Biofuels from microalgae—a review of technologies for production, processing, and extractions of biofuels and co-products. Renewable and Sustainable Energy Reviews 14: 557–577.

Bridgwater, A.V., D. Meier and D. Radlein. 1999. An overview of fast pyrolysis of biomass. Org. Geochem. 30: 1479–1493.

Bridgwater, A.V. and G.V.C. Peacocke. 2000. Fast pyrolysis processes for biomass. Renew Sustain Energy Rev. 4: 1–73.

Brunner, G. 2009. Near critical and supercritical water. Part I. Hydrolytic and hydrothermal processes. J. Supercrit. Fluids 47: 373–381.

Budarin, V.L., Y.Z. Zhao, M.J. Gronnow, P.S. Shuttleworth, S.W. Breeden, D.J. Macquarrie and J.H. Clark. 2011. Microwave-mediated pyrolysis of macro-algae. Green Chem. 13: 2330–2333.

Cantero, D.A., M.D. Bermejo and M.J. Cocero. 2015a. Governing chemistry of cellulose hydrolysis in supercritical water. Chem. Sus. Chem. 8: 1026–1033.

Cantero, D.A., Á.S. Tapia, M.D. Bermejo and M.J. Cocero. 2015b. Pressure and temperature effect on cellulose hydrolysis in pressurized water. Chem. Eng. J. 276: 145–154.

Cao, X., K.S. Ro, J.A. Libra, C.I. Kammann, I. Lima, N. Berge et al. 2013. Effects of biomass types and carbonization conditions on the chemical characteristics of hydrochars. J. Agric. Food Chem. 61: 9401–9411.

Castello, D., A. Kruse and L. Fiori. 2014. Supercritical water gasification of hydrochar. Chem. Eng. Res. Des. 92: 1864–1875.

Chakinala, A.G., D.W.F. Brilman, W.P.M. van Swaaij and S.R.A. Kersten. 2009. Catalytic and non-catalytic supercritical water gasification of microalgae and glycerol. Ind. Eng. Chem. Res. 49: 1113–1122.

Cherad, R., J.A. Onwudili, U. Ekpo, P.T. Williams, A.R. Lea-Langton, M. Carmargo-Valero and A.B. Ross. 2013. Macroalgae supercritical water gasification combined with nutrient recycling for microalgae cultivation. Environ. Prog. Sustainable Energy 32: 902–909.

Cherad, R., J.A. Onwudili, P.T. Williams and A.B. Ross. 2014. A parametric study on supercritical water gasification of *Laminaria hyperborea*: A carbohydrate-rich macroalga. Bioresour. Technol. 169: 573–580.

Cherad, R., J.A. Onwudili, P. Biller, P.T. Williams and A.B. Ross. 2016. Hydrogen production from the catalytic supercritical water gasification of process water generated from hydrothermal liquefaction of microalgae. Fuel 166: 24–28.

Chow, M.C., W.R. Jackson, A.L. Chaffee and M. Marshall. 2013. Thermal treatment of algae for production of biofuel. Energy Fuel 27: 1926–1950.

Chumpoo, J. and P. Prasassarakich. 2010. Bio-oil from hydro-liquefaction of bagasse in supercritical ethanol. Energy Fuels 24: 2071–2077.

Demirbaş, A. 2000. Effect of lignin content on aqueous liquefaction products of biomass. Energy Convers. Manag. 41: 1601–1607.

Demirbaş, A. 2005. Thermochemical conversion of biomass to liquid products in the aqueous medium. Energy Source 27: 1235–1243.

Demirbas, A. 2006. Oily products from mosses and algae via pyrolysis. Energy Sources, Part A: Recovery, Utilization, and Environmental Effects 28: 933–940.

Dinjus, E., A. Kruse and N. Troeger. 2011. Hydrothermal carbonization–1. Influence of lignin in lignocelluloses. Chem. Eng. Technol. 34: 2037–2043.

Dote, Y., S. Sawayama, S. Inoue, T. Minowa and S.Y. Yokoyama. 1994. Recovery of liquid fuel from hydrocarbon-rich microalgae by thermochemical liquefaction. Fuel 73: 1855–1857.

Douglas, C.E., Patrick Biller, Andrew B. Ross, Andrew J. Schmidt and S.B. Jones. 2015. Hydrothermal liquefaction of biomass: Developments from batch to continuous process. Bioresource Technology 178: 147–156.

Du, Z.Y., Y.C. Li, X.Q. Wang, Y.Q. Wan, Q. Chen, C.G. Wang et al. 2011. Microwave-assisted pyrolysis of microalgae for biofuel production. Bioresour. Technol. 102: 4890–4896.

Ekpo, U., A.B. Ross, M.A. Camargo-Valero and P.T. Williams. 2016. A comparison of product yields and inorganic content in process streams following thermal hydrolysis and hydrothermal processing of microalgae, manure and digestate. Bioresource Technology 200: 951–960.

Elliott, D.C., T.R. Hart, A.J. Schmidt, G.G. Neuenschwander, L.J. Rotness, M.V. Olarte et al. 2013. Process development for hydrothermal liquefaction of algae feedstocks in a continuous-flow reactor. Algal Research 2: 445–454.

Elliott, D.C., P. Biller, A. Ross, A.J. Schmidt and S.B. Jones. 2015. Hydrothermal liquefaction of biomass: Developments from batch to continuous process. Bioresource Technology 178: 147–156.

Elliott, D.C. 2016. Review of recent reports on process technology for thermochemical conversion of whole algae to liquid fuels. Algal Research 13: 255–263.

Falco, C., N. Baccile and M.-M. Titirici. 2011. Morphological and structural differences between glucose, cellulose and lignocellulosic biomass derived hydrothermal carbons. Green Chem. 13: 3273–3281.

Fields, M.W., A. Hise, E.J. Lohman, T. Bell, R.D. Gardner, L. Corredor et al. 2014. Sources and resources: Importance of nutrients, resource allocation, and ecology in microalgal cultivation for lipid accumulation. Appl. Microbiol. Biotechnol. 98: 4805–4816.

Funke, A. and F. Ziegler. 2010. Hydrothermal carbonization of biomass: A summary and discussion of chemical mechanisms for process engineering: A review. Biofuels Bioprod. Bioref. 4: 160–177.

Goodwin, A.K. and G.L. Rorrer. 2009. Conversion of xylose and xylosephenol mixtures to hydrogen rich gas by supercritical water in an isothermal microtube flow reactor. Energy & Fuels 23: 3818–3825.

Goudriaan, F.D. and G.R. Peferoen. 1990. Liquid fuels from biomass via a hydrothermal process. Chemical Engineering Science 45: 2729–2734.

Graz, Y., S. Bostyn, T. Richard, P. Escot Bocanegra, E. de Bilbao, J. Poirier et al. 2016. Hydrothermal conversion of *Ulva* macro algae in supercritical water. J. of Supercritical Fluids 107: 182–188.

He, C., A. Giannis and J.-Y. Wang. 2013. Conversion of sewage sludge to clean solid fuel using hydrothermal carbonization: Hydrochar fuel characteristics and combustion behavior. Applied Energy 111: 257–266.

Heilmann, S.M., H.T. Davis, L.R. Jader, P.A. Lefebvre, M.J. Sadowsky, F.J. Schendel et al. 2010. Hydrothermal carbonization of microalgae. Biomass and Bioenergy 34: 875–882.

Heilmann, S.M., L.R. Jader, L.A. Harned, M.J. Sadowsky, F.J. Schendel, P.A. Lefebvre et al. 2011. Hydrothermal carbonization of microalgae II. Fatty acid, char, and algal nutrient products. Applied Energy 88: 3286–3290.

Hirano, A., K. Hon-Nami, S. Kunito, M. Hada and Y. Ogushi. 1998. Temperature effect on continuous gasification of microalgal biomass: theoretical yield of methanol production and its energy balance. Catalysis Today 45: 399–404.

Hoekman, S.K., A. Broch and C. Robbins. 2011. Hydrothermal carbonization (HTC) of lignocellulosic biomass. Energy & Fuels 25: 1802–1810.

Hu, Z.Q., Y. Zheng, F. Yan, B. Xiao and S.M. Liu. 2013. Bio-oil production through pyrolysis of blue-green algae blooms (BGAB): Product distribution and bio-oil characterization. Energy 52: 119–125.

Huang, G., F. Chen, D. Wei, X.W. Zhang and G. Chen. 2010. Biodiesel production by microalgal biotechnology. Applied Energy 87: 38–46.

Jain, A., R. Balasubramanian and M.P. Srinivasan. 2016. Hydrothermal conversion of biomass waste to activated carbon with high porosity: A review. Chemical Engineering Journal 283: 789–805.

Jena, U. and K.C. Das. 2011. Comparative evaluation of thermochemical liquefaction and pyrolysis for bio-oil production from microalgae. Energy & Fuels 25: 5472–5482.

Jindal, M.K. and M.K. Jha. 2015. Effect of process conditions on hydrothermal liquefaction of biomass. IJCBS Research Paper 2(8): 180–188.

Kang, S., X. Li, J. Fan and J. Chang. 2012. Characterization of hydrochars produced by hydrothermal carbonization of lignin, cellulose, D-xylose, and wood meal. Ind. Eng. Chem. Res. 51: 9023–9031.

Karagöz, S., T. Bhaskar, A. Muto and Y. Sakata. 2005. Comparative studies of oil compositions produced from sawdust, rice husk, lignin and cellulose by hydrothermal treatment. Fuel 84: 875–884.

Knezevic, D., W. van Swaaij and S. Kersten. 2009. Hydrothermal conversion of biomass: I. glucose conversion in hot compressed water. Ind. Eng. Chem. Res. 48: 4731–4743.

Lee, A., D. Lewis, T. Kalaitzidis and P. Ashman. 2016. Technical issues in the large-scale hydrothermal liquefaction of microalgae biomass to biocrude. Current Opinion in Biotechnology 38: 85–89.

Libra, J.A., K.S. Ro, C. Kammann, A. Funke, N.D. Berge, Y. Neubauer et al. 2011a. Hydrothermal carbonization of biomass residuals: a comparative review of the chemistry, processes and applications of wet and dry pyrolysis. Biofuels 2: 71–106.

Libra, J.A., S.R. Kyoung, C. Kammann, A. Funke, N.D. Berge, Y. Neubauer et al. 2011b. Hydrothermal carbonization of biomass residuals: A comparative review of the chemistry, processes and applications of wet and dry pyrolysis. Biofuels 2: 89–124.

Liu, X., B. Saydeh, P. Ernki, L.M. Colosi, M.B. Greg, J. Rhoses and A.F. Clarens. 2013. Pilot-scale data provide enhanced estimates of the life cycle energy and emissions profile of algae biofuels produced via hydrothermal liquefaction. Bioresource Technology 148: 163–171.

López, D., W. Prins, F. Ronsse and W. Brilman. 2013. Hydrothermal liquefaction (HTL) of microalgae for biofuel production: State of the art review and future prospects. Biomass Bioenergy 53: 113–127.

Lu, Y., L. Zhao, Q. Han, L. Wei, X. Zhang and L. Guo et al. 2013. Minimum fluidization velocities for supercritical water fluidized bed within the range of 633–693 K and 23–27 MPa. Int. J. Multiph. Flow 49: 78–82.

Mani, S., L.G. Tabil and S. Sokhansanj. 2004. Grinding performance and physical properties of wheat and barley straws, corn stover, and switchgrass. Biomass Bioenergy 27: 339–352.

Marcilla, A., L. Catalá, J.C. García-Quesada, F.J. Valdés and M.R. Hernández. 2013. A review of thermochemical conversion of microalgae. Renew. Sustain. Energy Rev. 27: 11–19.

Marsman, J., J. Wildschut, F. Mahfud and H. Heeres. 2007. Identification of components in fast pyrolysis oil and upgraded products by comprehensive two dimensional gas chromatography and flame ionisation detection. J. Chromatogr. A 1150: 21–27.

Mazaheri, H., K.T. Lee, S. Bhatia and A.R. Mohamed. 2010. Subcritical water liquefaction of oil palm fruit press fiber for the production of bio-oil: Effect of catalysts. Bioresource Technology 101: 745–751.

Miao, X.L. and Q.Y. Wu. 2004a. Fast pyrolysis of microalgae to produce renewable fuels. J. Anal. Appl. Pyrol. 71: 855–863.

Miao, X.L. and Q.Y. Wu. 2004b. High yield bio-oil production from fast pyrolysis by metabolic controlling of *Chlorella prototothecoides*. J. Biotechnol. 110: 85–93.

Milledge, J.J., B. Smith, P.W. Dyer and P. Harvey. 2014. Macroalgae-derived biofuel: A review of methods of energy extraction from seaweed biomass. Energies 7: 7194–7222.

Minowa, T., S.Y. Yokoyama, M. Kishimoto and T. Okakura. 1995. Oil production from algal cells of *Dunaliella tertiolecta* by direct thermochemical liquefaction. Fuel 74: 1735–1738.

Minowa, T., T. Kondo and S.T. Sudirjo. 1998. Thermochemical liquefaction of Indonesian biomass residues. Biomass Bioenergy 14: 517–524.

Minowa, T. and S. Sawayama. 1999. A novel microalgal system for energy production with nitrogen cycling. Fuel 78: 1213–1215.

Nakada, S., D. Saygin and D. Gielen. 2014. Global bioenergy supply and demand projections. International Renewable Energy Agency. Abu Dhabi-UAE.

Neveux, N., A.K.L. Yuen, C. Jazrawi, M. Magnusson, B.S. Haynes, A.F. Masters, A. Montoya, N.A. Paul, T. Maschmeyer and R. de Nys. 2014. Biocrude yield and productivity from the hydrothermal liquefaction of marine and freshwater green macroalgae. Bioresource Technology 155: 334–341.

Ozkan, A., K. Kinney, L. Katz and H. Berberoglu. 2012. Reduction of water and energy requirement of algae cultivation using an algae biofilm photobioreactor. Bioresource Technology 114: 542–548.

Öztürk, I., S. Irmak, A. Hesenov and O. Erbatur. 2010. Hydrolysis of kenaf (*Hibiscus cannabinus* L.) stems by catalytical thermal treatment in subcritical water. Biomass Bioenergy 34: 1578–1585.

Pan, P., C.W. Hu, W.Y. Yang, Y.S. Li, L.L. Dong, L.F. Zhu, D. Tong, R. Qing and Y. Fan. 2010. The direct pyrolysis and catalytic pyrolysis of *Nannochloropsis* sp. residue for renewable bio oils. Bioresour. Technol. 101: 4593–4599.

Patel, B., M. Guo, A. Izadpanah, N. Shah and K. Hellgardt. 2016. A review on hydrothermal pre-treatment technologies and environmental profiles of algal biomass processing. Bioresource Technology 199: 288–299.

Peng, W.M. and Q.Y. Wu. 2000. Effects of temperature and holding time on production of renewable fuels from pyrolysis of *Chlorella protothecoides*. J. Appl. Phycol. 12: 147–52.

Peterson, A.A., F. Vogel, R.P. Lachance, M. Froling, M.J. Antal, Jr. and J.W. Tester. 2008. Thermochemical biofuel production in hydrothermal media: A review of sub- and supercritical water technologies. Energy Environ. Sci. 1: 32–65.

Prado, J.M., D. Lachos-Perez, T. Forster-Carneiro and M.A. Rostagno. 2016. Sub- and supercritical water hydrolysis of agricultural and food industry residues for the production of fermentable sugars: A review. Food and Bioproducts Processing 98: 95–123.

Reddy, S.N., S. Nanda, A.K. Dalai and J.A. Kozinski. 2014. Supercritical water gasification of biomass for hydrogen production. International Journal of Hydrogen Energy 39: 6912–6926.

Romero-Anaya, A., M. Ouzzine, M. Lillo-Ródenas and A. Linares-Solano. 2014. Spherical carbons: Synthesis, characterization and activation processes. Carbon 68: 296–307.

Saber, M., B. Nakhshiniev and K. Yoshikawa. 2016. A review of production and upgrading of algal bio-oil. Renewable and Sustainable Energy Reviews 58: 918–930.

Schacht, C., C. Zetzl and G. Brunner. 2008. From plant materials to ethanol by means of supercritical fluid technology. J. Supercrit. Fluids 46: 299–321.

Schnurr, P.J., G.S. Espie and D.G. Allen. 2013. Algae biofilm growth and the potential to stimulate lipid accumulation through nutrient starvation. Bioresource Technology 136: 337–344.

Sevilla, M. and A. Fuertes. 2009a. Chemical and structural properties of carbonaceous products obtained by hydrothermal carbonization of saccharides. Chem. Eur. J. 15: 4195–4203.

Sevilla, M. and A. Fuertes. 2009b. The production of carbon materials by hydrothermal carbonization of cellulose. Carbon 47: 2281–2289.

Shekh, A.Y., P. Shrivastava, K. Krishnamurthi, S.N. Mudliar, S.S. Devi, G.S. Kanade et al. 2013. Stress-induced lipids are unsuitable as a direct biodiesel feedstock: A case study with *Chlorella pyrenoidosa*. Bioresource Technology 138: 382–386.

Shijie, L., J. Witter, D. Vardon, B. Sharma, J. Guesta and T. Strathmann. 2015. Prediction of microalgae hydrothermal liquefaction products from feedstock biochemical composition. Green Chemistry 6: 3584–3599.

Shuping, Z., Y. Wu, M. Yang, C. Li and J. Tong. 2010. Bio-oil production from sub- and supercritical water liquefaction of microalgae *Dunaliella tertiolecta* and related properties. Energy Environ. Sci. 3: 1073–1078.

Stucki, S., F. Vogel, C. Ludwig, A.G. Haiduc and M. Brandenberger. 2009. Catalytic gasification of algae in supercritical water for biofuel production and carbon capture. Energy Environ. Sci. 2: 535–541.

Sun, Y.N., B. Gao, Y. Yao, J.N. Fang, M. Zhang, Y.M. Zhou et al. 2014. Effects of feedstock type, production method, and pyrolysis temperature on biochar and hydrochar properties. Chem. Eng. J. 240: 574–578.

Tekin, K., S. Karagöz and S. Bektaş. 2014. A review of hydrothermal biomass processing. Renewable and Sustainable Energy Reviews 40: 673–687.

Tiong, L., M. Komiyama, Y. Uemura and T.T. Nguyen. 2016. Catalytic supercritical water gasification of microalgae: Comparison of Chlorella vulgaris and *Scenedesmus quadricauda*. J. of Supercritical Fluids 107: 408–413.

Titirici, M.-M., R.J. White, C. Falco and M. Sevilla. 2012. Black perspectives for a green future: hydrothermal carbons for environment protection and energy storage. Energy Environ. Sci. 5: 6796–6822.

Toor, S.S., H. Reddy, S. Deng, J. Hoffmann, D. Spangsmark, L.B. Madsen et al. 2013. Hydrothermal liquefaction of *Spirulina* and *Nannochloropsis salina* under subcritical and supercritical water conditions. Bioresource Technology 131: 413–419.

Vardon, D.R., B.K. Sharma, G.V. Blazina, K. Rajagopalan and T.J. Strathmann. 2012. Thermochemical conversion of raw and defatted algal biomass via hydrothermal liquefaction and slow pyrolysis. Bioresource Technology 109: 178–187.

Wagner, J., R. Bransgrove, T.A. Beacham, M.J. Allen, K. Meixner, B. Drosg et al. 2016. Co-production of bio-oil and propylene through the hydrothermal liquefaction of polyhydroxybutyrate producing cyanobacteria. Bioresource Technology 207: 166–174.

Wang, K.G., R.C. Brown, S. Homsy, L. Martinez and S.S. Sidhu. 2013. Fast pyrolysis of microalgae remnants in a fluidized bed reactor for bio-oil and biochar production. Bioresour. Technol. 127: 494–9.

Watanabe, Y., H. Abe, Y. Daigo, R. Fujisawa and M. Sakaihara. 2004. Effect of physical property and chemistry of water on cracking of stainless steels in sub-critical and supercritical water. Key Engineering Materials. 261-263: 1031–1036.

Williams, P.T. and J. Onwudili. 2005. Composition of products from the supercritical water gasification of glucose: A model biomass compound. Ind. Eng. Chem. Res. 44: 8739–8749.

Yakaboylu, O., J. Harinck, K.G. Smit and W. de Jong. 2015. Supercritical water gasification of biomass: a detailed process modelling analysis for a microalgae gasification process. Ind. Eng. Chem. Res. 54: 5550–5562.

Yang, X., F. Guo, S. Xue and X. Wang. 2016. Carbon distribution of algae-based alternative aviation fuel obtained by different pathways. Renewable and Sustainable Energy Reviews 54: 1129–1147.

Yildiz, G., F. Ronsse, R. Duren and W. Prins. 2016. Challenges in the design and operation of processes for catalytic fast pyrolysis of woody biomass. Renewable and Sustainable Energy Reviews 57: 1596–1610.

Yusuf, C. 2013. Constrains to commercialization of algal fuels. Journal of Biotechnology 167: 201–214.

Zhang, B., M. von Keitz and K. Valentas. 2009. Thermochemical liquefaction of high-diversity grassland perennials. Journal of Analytical and Applied Pyrolysis 84: 18–24.

Zhang, J.-H., Q.-M. Lin and X.-R. Zhao. 2014. The hydrochar characters of municipal sewage sludge under different hydrothermal temperatures and durations. J. Integr. Agric. 13: 471–482.

Zhoufan, C., D. Peigao and X. Yuping. 2015. Catalytic hydropyrolysis of microalgae: Influence of operating variables on the formation and composition of bio-oil. Bioresource Technology 184: 349–354.

CHAPTER 6

Ethanol Production from Macroalgae Biomass

*Jorge M.T.B. Varejão** and *Raphaela Nazaré*

1. Introduction

Humanity continues to depend mainly on fossil fuels as the main source of energy and raw materials. A standard of life quality is accessible to an increasing part of the population, which demands for growing energy supply. At present, most of it comes from the burning and refining of fossil fuels. The decrease of this dependence is an urgent goal, as the discoveries of new important petrol reserves in the subsoil are becoming scarce (Miller and Sorrell 2014), and its combustion emissions are raising atmospheric carbon dioxide content. This has a big contribution to make to the greenhouse effect, which starts to make big environmental damages to the earth. In the long term, continuing to extract fossil fuels from the soil seems unfeasible, due to these earth-damaging effects. Long-term energy supply for the society must rely on renewable sources; solar energy and biomass in its different variants are the most obvious alternatives. Biomass is rich in fat materials and/or carbohydrates, which can be converted into liquid biofuels. It is expected that the electric automotive substitution along the century, however conventional, will last for a long time. Liquid fuels, due to their high-density energy content, should be difficult to substitute in particular uses, such as commercial flight, and their use as solvents and disinfectants may even increase. If lipid materials are available, its conversion into biodiesel is a convenient solution, since the diesel engine is widely used (GuanHua et al. 2010). In the case of carbohydrate-rich substrates, conversion in bioethanol

Instituto Politécnico de Coimbra, Escola Superior Agrária, CERNAS, Department of Exact Science, Bencanta. 3000-316 Coimbra. Portugal.
* Corresponding author: jvarejao@esac.pt

is the best solution, capable of being used in a gasoline automobile motor, mixed with gasoline, and even in its pure form as it occurs in Brazil. Both approaches have neutral emissions contribution (Hill et al. 2006).

The renewable bioethanol fuel used at present originated from agricultural and forest biomass. Brazil is a major ethanol producer through the fermentation of sugar cane molasses waste (55% from total sugar converted in 2010). In Canada and USA low quality cereal is transformed in ethanol (in 2010 40% from total corn production was converted) (Enquist-Newman et al. 2014). However, these sources can only produce part of the required amount of biofuels and make strong competition with food production resources and land use and depletion. By 2050, an increase in food production of 70% relative to the level of 2005 is needed (FAO 2009) to satisfy the demand of an increasing population. By the same time, the increase in land available for agriculture is only expected to rise about 5% (FAO 2009). Finding others sources of energy is an urgent goal, and more research into alternatives are needed. The lignocellulosic materials, mostly considered waste materials derived from agricultural and forest activities, do not compete with food and land use, and their conversion is very promising. In fact, a big effort has been made in the last few decades to try to achieve the technological means to achieve its competitive conversion into biofuels (Srivastava et al. 2015). However, technology demonstrated that it is difficult to convert the carbohydrate in lignocellulosic materials in biofuels, mainly due to its intricate and strong structure, where the high cellulose crystallinity and the presence of lignin (Chapple et al. 2007) are two main problems. Both confer to efficient conversion the need of different physical and chemical treatments, posing difficulties to producers in the competitiveness of cellulosic ethanol.

More recently, an alternate possibility was brought to study, and refers to the use of algae biomass as a source of sugars or fat materials to the bio-refinery purpose (Wei et al. 2013). Both microalgae or macroalgae are rich in carbohydrates, have elevated productivity per area, fast growth rates (Lee et al. 2013), do not compete with land for food, do not require fresh water supply, and potentially are easier to convert in final products than lignocellulosic biomass (Trivedi et al. 2015). Microalgae can be rich in lipids, and may constitute a source of biodiesel (Guedes et al. 2011). Macroalgae have generally very low lipid content, however, their dry mass is rich in carbohydrates reaching values as high as 60–70%, making them a possible source for bioethanol (Huang et al. 2010, Mata et al. 2010, Kraan 2012, Daroch et al. 2013, Trivedi et al. 2015, Chen et al. 2015).

2. Macroalgae Species and Biomass Composition

Sweet water and marine algae are known to have very high biomass productivity per annum, and thus may function as a very good carbon dioxide sink. Microalgae species may have biomass productivity ranging from 1.0 to 3.4 kg carbon m⁻².year⁻¹ (Kraan 2013). Macroalgae productivity can reach values up to 13.0 kg m⁻².year⁻¹ on dry weight basis for some species (Fleurence and Levine 2016). Both are considered to have at least three times more biomass productivity than the highest obtained by any land plant, in which sugar cane is considered to be one of the most productive, with up to 10 kg.m⁻².year⁻¹ in raw basis biomass, where values are expressed in dry weight of 6–8 kg.m⁻².year⁻¹.

Several kelp species can be used, and these are related to its regional availability. Macroalgae are generally divided in brown, red and green algae. In Table 6.1, some species being considered in the study are presented together with their main chemical composition constituents.

As can be seen in Table 6.1, macroalgae are rich in carbohydrate with a content expressed in a dry matter in the range of 40–65%. A remarkable difference between macroalgae biomass and terrestrial vegetable is the near absence of lignin; in the range of 0–3.0% (not shown in Table 6.1), a component often referred to as being responsible for some of the difficulty of saccharification of carbohydrates polymers in plants (Chapple et al. 2007). Lipid content attains very low values, which in principle indicate that fuels derived from lipids, such as biodiesel, are not feasible, at least with these species of macroalgae. On the contrary, their conversion to bioethanol, attributing to its high content of carbohydrate polymers seems a possibility. The structure of the more significant ones has been already fully characterized.

Figure 6.1 illustrates the relevant structural aspects of main polysaccharides found in macroalgae. Red macroalgae are rich in agar (1) which main monomer is a disaccharide of β(1,4) linked D-galactose to an 3,6-anhydro-L-galactopyranose; carrageenan whose structure shows different types of sulphated saccharides alternated with β- and α-galactose; examples (2) and (3) are shown in Fig. 6.1. Cellulose may also be present (not shown), a β(1,4)-D-glucose linear polymer.

Brown algae identified as polysaccharides include alginic acid: a linear polymer with 1,4-linked β-D-mannuronic acid (M) and α-L-guluronic acid (G) in varying sequence; laminarin whose main units are β(1,3) glucose saccharides with a few β(1,6) linkages; fucoidan-mainly 1,2-linked α-L-fucose-4-sulphate units, see structure (6) in Fig. 6.1 and cellulose.

Green macroalgae have composition rich in starch and cellulose (Wei et al. 2013). The main differences between macroalgae and terrestrial plants monosaccharides are the presence of oses with oxidation of hexose sixth carbon—the uronic acids—a good example of which is alginic acid with mannuronic and galacturonic acids sub units (see 5 and 6, Fig. 6.1), sugars that simultaneously contain the aldehyde and a carboxylic group (Sánchez-Machado et al. 2004). In general, macroalgae have a higher richness in glycosidic linkages types in terms of hexose carbon number, both in alfa

Table 6.1. Biomass water content and main chemical composition of dry matter for examples of red, brown, and green macroalgae.

Macroalgae	Order	Genera	Water (%)	Lipid (%)	Protein (%)	Ash (%)	Polysaccharides (%)	Ref.
Red (Rhodophyta)	Gracilariales	*Gracilaria tenuistipitata*	89.13	0.26	6.11	22.91	59.90	Chirapart et al. 2014
	Gelidiales	*Gelidium amansii*	74.40	-	18.30	7.40	-	Jang et al. 2012
	Gracilariales	*Gracilaria gracilis*	92.01	0.06	50.00	24.8	46.60	Rodrigues et al. 2015
Brown (Phaeophyceae)	Laminariales	*Saccharina japonica*		1.60	10.60	21.80	66.00	Jang et al. 2012
		Saccharina japonica		1.5–1.8	8.0–15.0	31.0–32.0	52.0–60.0	Murphy et al. 2013
	Fucales	*Sargassum muticum*	67.40	1.45	16.90	22.94	49.30	Rodrigues et al. 2015
		Sargassum fulvellum		0.50	19.90	35.10	48.00	Jang et al. 2012
		Sargassum fulvellum		1.40	13.00	46.00	39.60	Murphy et al. 2013
	Laminariales	*Undaria pinnatifida*		1.80	18.30	28.00	55.60	Jang et al. 2012
	Fucales	*Sargassum fusiforme* (formerly *Hizikia fusiformis*)		0.40	13.90	26.60	63.20	Jang et al. 2012
Green (Chlorophyta)	Ulvales	*Ulva linza* (formerly *Enteromorpha linza*)		1.80	31.60	29.20	39.80	Jang et al. 2012
		Ulva instestinalis	86.70	0.62	10.59	20.65	53.8	Chirapart et al. 2014

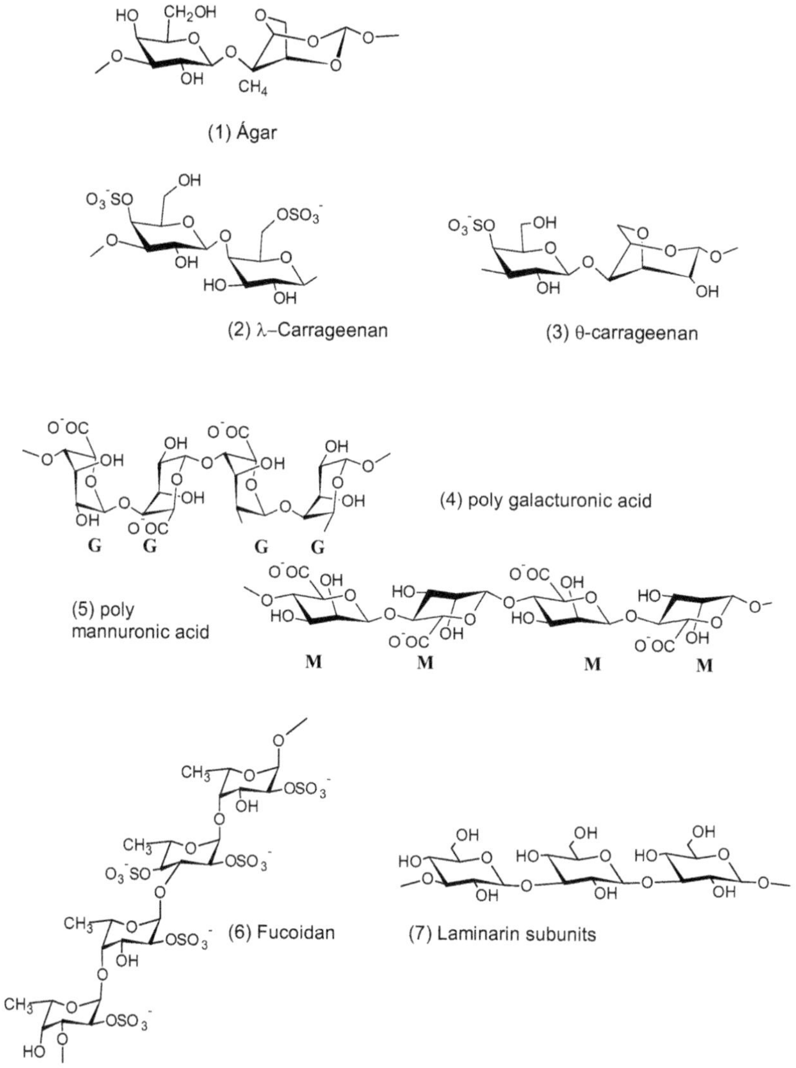

Figure 6.1. Relevant saccharide structure subunits found in different type of macroalgae polysaccharides.

and beta forms. Concurrently, anhydrous sugars are present in agar and in carrageenan together with the presence of sulphated sugars.

Fucoidan and carrageenan sulphated degree may be inferred by the elemental content of sulfur. In Table 6.2, the levels of sulfate in some macroalgae are presented.

Table 6.2. Sulfate content expressed in percentage of dry weight (% dw) in tissue of the different seaweeds.

Algal species	Sulfate content (%)[a]
Ulva intestinalis (Chlorophyta)	8.24 ± 0.28
Rhizoclonium riparium (Chlorophyta)	1.97 ± 0.20
Gracilaria salicornia (Rhodophyta)	4.69 ± 0.04
Gracilaria tenuistipitata (Rhodophyta)	8.58 ± 0.36

[a.] Dry weight percentage. Adapted from Chirapart et al. 2014.

After saccharification, common HPLC and GC techniques are more amenable to identify the sugars released and knowledge of different polymer carbohydrates comes through this indirect evidence. The problem is that the precise way of connection can be lost in saccharification, and must be obtained by techniques such as IR spectroscopy.

3. Macroalgae Polysaccharide Saccharification

3.1 *Acid Treatment followed by Enzymatic Degradation*

Most of the studies involved in the saccharification of macroalgae biomass rely on a first step of diluted acid or base heat treatment, followed by an enzymatic hydrolysis with beta glucosidases (Wei et al. 2013). In some cases, reports exist for the use of amylase enzymes (Daroch et al. 2013). The idea behind this strategy is that α-(1,4) glycosidic linkage is known to be very easily broken, as it occurs in starch materials products in which cooking with just hot water is enough to break the bond. Glycosidic α-(1,3) linkages occur in laminarin, and are also very easy to break with acid/heat (Khambhaty et al. 2012). If the biomass contains sulphate glucan as in carrageenan, the same acid/heat process will hydrolyze the sulphate group, yielding a no sulphate sugar, galactose for example (Khambhaty et al. 2012). This led to the theory that the diluted acid/heat treatment makes sense in saccharifying macroalgae biomass. The more recalcitrant glycosidic linkage that can resist to the acid/heat treatment is the β-(1-4) connection, known to be more difficult to break with the acid hydrolysis. A vast background of knowledge is available from the cellulose saccharification studies. In order to circumvent this difficulty, β-glucosidase enzymes are chosen to be used. The use of both treatments in a sequential order should allow a good saccharification yield from initial polysaccharides.

The experimental conditions usually considered the use of diluted acid- most of the times sulfuric acid in the concentration range of 0.2 to 5.0%, and with temperature in the range of 100 to 210°C (Lee et al. 2013). The use of higher temperatures while yielding good saccharification yields may suffer a drawback, which is the conversion of already saccharified

C5 and C6 sugars to furfural and 2-hydroximethylfurfural (HMF). These substances, even in low concentrations, are well known to be strong yeast growth inhibitors, making necessary intermediate detoxification steps in the conversion of sugars to ethanol (Liu et al. 2004). However, when looking at macroalgae carbohydrate polymers, a less intricate and non-crystalline structure is evident with the presence of both α and β sugar linkage. This suggests that more amenable conditions may be used.

3.2 Acid Hydrolysis

The study of lignocellulosic materials suggest that the beta-glycosidic bond can be hydrolised with diluted acid/heat, being the major problem its inacessibility due to crystallinity of cellulose. In macroalgae, rigidity is not as necessary as in higher plants, making the polysaccharide polymers much easier to break using simple acid/heat treatment, even if they have the beta linkage present. The biomass of the red macroalgae *Kappaphycus alvarezii* yields on simple acid saccharification with 2,5% sulfuric acid at 100°C, yields after fermentation is as high as 90–94% based on carbohydrate mass content (Khambhaty et al. 2012). In this case, no use of expensive enzymes was necessary. Results obtained in Coimbra (Nazaré 2015) with *Sargassum* spp. suggest the same behavior, use of sulfuric acid is very effective in saccharifying macroalgae biomass (see below).

3.3 Use of Enzymes

Alginate in natural systems can be depolymerized by endo and exo type alginate lyases (Wang et al. 2014); other saccharides lyases may exist and need to be recognized from microorganisms which depolymerize them, and more study is needed in identifying such species (Jang et al. 2012). This knowledge with genetic reengineering techniques may possibly turn modified yeast that allows direct alginate fermentation to ethanol.

4. Fermentation of Macroalgae Hydrolysates

Most macroalgae show a carbohydrate polymers content relatively to dry mass in the range of 60–70% (Gorham and Lewey 1984). Considering that all the polymers can be saccharified to monomers and that these are fermentable by common yeast such as *Saccharomyces cerevisiae*, a conversion of 45% carbohydrate is expected to yield ethanol, which means 27 g ethanol by 100 g dry macroalgae, or by volume a 40 mL ethanol from 100 g dry biomass. In Table 6.3, the ethanol production from macroalgae biomass, published in the last few years, together with the conditions used to obtain them is presented.

Table 6.3. Survey in literature for saccharification yield/ethanol production from macroalgae biomass.

Weed	Load % (w/v)	Treatments	Process	Saccharification yield (%)	Microorganism	Ethanol conversion (%)[a]	Reference
Ulva fasciata		Sodium acetate, 120°C, 1 h	Celluclast 22119, 45°C, 36 h	40.0	*Saccharomyces cerevisiae*, MTCC 180	88.0	Delile et al. 2013
		HCl 0,1 to 1 M, 95°C, 15 h		3- to 98.0%		very low	Chirapart et al. 2014
Saccharina japonica	10	40 mM H_2SO_4, 121°C, 60 min		21.0			Jang et al. 2012
	10	1.5 KU/mL α-amilase Termamyl 120L, 40 mM H_2SO_4 121°C, 60 min		31.2			Jang et al. 2012
	10	40 mM H_2SO_4, 121°C, 60 min, 1 g dew/L *Bacillus* sp. JS-1 30°C, 200 rpm, 7.5 days	SSF	45.6	*Bacilus* spp.	33.3	Jang et al. 2012
Kappaphycus alvarezii		0.9 N H_2SO_4 at 120°C for 60 min	SHF		*Saccharomyces cerevisiae* NCIM 3455	92.3	Daroch et al. 2013
Gelidium elegans		Meicelase treatment 50°C for 120 h pH 5.5, 48 h	SHF		*Saccharomyces cerevisiae* IAM 4178	36.7	Yanagisawa et al. 2011
Sargassum spp.		H_2SO_4, 1.0–6.4%, 115–130°C, 0.1–2 h	SHF		*Saccharomyces cerevisiae*	64.0–89.0	Borines et al. 2013, Myra et al. 2013

Sargassum multicum	2.5	H_2SO_4, 2.5–5%, 100°C, 1 h	SHF[b]	15.0–58.0%	*Saccharomyces cerevisiae*	28.0–96.0	Nazaré 2015
	2.5	Cellulase 50 FPU/g, pH = 5, 40°C, 1 week	SHF[b]	33.9	*Saccharomyces cerevisiae*	32.6	Nazaré 2015
	2.5	H_2SO_4, 4%, 100°C, 1 h	SHF[b]	15–58%	*Saccharomyces cerevisiae*	96.2	Nazaré 2015

[a]. Relatively to the dry biomass weight; [b]. Non septic conditions

The macroalgal biomass usually require drying to some extent and grinding. After these physical operations, mainly two processes are used: a preliminary sacharification of macroalgae biomass involving or no several pretreatments, after which the fermentation is made. This process is known as Separated Hydrolysis and Fermentation (SHF). Especially when using enzymes, this process may be less recommended, since enzymes are often regulated and inhibited by the presence of sugars. In this case, it may be better to make the so called Simultaneously Saccharification and Fermentation (SSF).

Daroch et al. (2013) collected some work of bioethanol production by pretreatment and fermentation of micro and macroalgae biomass, reporting yields in ethanol in the range of 0.47 to 40.0% yield. This means that in certain cases, the yield is close to the theoretical maximum.

Borines et al. (2013) used a sequential sulfuric acid hydrolysis followed by cellulase enzyme treatment obtaining from *Sargassum* sp. biomass ethanol yields as high as 89.1% after *Saccharomyces cerevisiae* fermentation. In Borines' words: "The ethanol conversions obtained were markedly higher than the theoretical yield based on glucose as substrate". Results pointing for an almost sugar conversion to ethanol besides the sugar structural variety was also made using SSF reports of *Saccharina japonica* biomass conversion in ethanol with a yield of 67.0% (Lee et al. 2013).

Macroalgae saccharified broths contain a much diversified range of sugars, whose fermentability by *Saccharomyces cerevisiae* or others microorganisms was not yet studied in most cases. Use of *Saccharomyces cerevisae* is desirable, since it is viable up to concentrations of ethanol of 12–14% (v/v), contents of which are required for successful separation of ethanol with high purity. Uronic acids are referred to be not fermentable by yeast, but not many studies have been published in the literature on the subject (Kraan 2012). In natural systems, endo and exo alginate lyases saccharify the alginate, which is then converted to 4-deoxy-L-erythro-5-hexoseulose uronic acid (DEH). This is the key intermediate considered not to be fermented by general yeasts (Wang et al. 2014). Genetic engineering techniques allowed this intermediate to suffer glycolysis with incorporation in *Saccharomyces cerevisiae* of enzymes from *Asteromyces cruciatus* (Wei et al. 2013). It is also known that bacteria can convert uronic acids to pyruvate and glyceraldeheyde-3-Phosphate, which can be fermented to ethanol by the glycolysis pathway (Wei et al. 2013, van Maris et al. 2006), suggesting that fermentation of hydrolysates from macroalgae can efficiently be converted to ethanol in aseptic conditions yeast fermentation. Results reinforcing this synergetic activity are obtained by Nazaré (Nazaré 2015), in which fermentation of sulfuric acid hydrolysates from *Sargassum* spp. with *Saccaromyces cerevisae*, under non septic conditions resulted in yields on ethanol conversion in excess of 95% (see Table 6.3).

5. Conclusion

Very promising results have already obtained in converting macroalgae biomass in bioethanol. Their conversion does not suffer from the difficulties found in the conversion of lignocellulosic materials, and has comparative environmental advantages to those materials. The field is in its initial stages of development and an extension of optimization will be possible upon research. This can support the idea that macroalgae will be a substrate to consider in biofuel production, meaning that efforts for its pharming and productivity growth are recommended to be made.

Keywords: Macroalgae; saccharification; bioethanol; fermentation; biofuels

References

Borines, M.G., R.L. de Leon and J.L. Cuello. 2013. Bioethanol production from the macroalgae *Sargassum* spp. Bioresource Technology 138: 22–29.

Chapple, C., M. Ladisch and R. Meilan. 2007. Loosening lignin's grip on biofuel production. Nature Biotechnology 25: 746–748.

Chen, H., D. Zhou, G.N. Luo, S. Zhang, J. Chen and J. Key. 2015. Macroalgae for biofuels production: progress and perspectives. Renewable and Sustainable Energy Reviews 47: 427–437.

Chirapart, A., A. Praiboon, P.S. Pongsatorn, C. Pattanapon and N. Ritnunraksa. 2014. Chemical composition and ethanol production potential of thai seaweed species. J. Appl. Phycol. 26: 979–986.

Daroch, M., S. Geng and G. Wang. 2013. Recent advances in liquid biofuel production from algal. Applied Energy 102: 1371–1381.

Delile, N.T., G. Vishal and C.R.K. Bhavanathj. 2013. Enzymatic hydrolysis and production of bioethanol from common macrophytic green alga *Ulva fasciata*. Bioresource Technology 150: 106–112.

Enquist-Newman, M., A.M. Faust, D.D. Bravo, C.N. Santos, R.M. Raisner, A. Hanel et al. 2014. Efficient ethanol production from brown macroalgae sugars by a synthetic yeast platform. Nature 505(7482): 239–243.

FAO. 2009. How to feed the world in 2050. Food and agriculture organization of United Nations. Available online at: http://www.fao.org/fileadmin/templates/wsfs/docs/expert_paper/How_to_Feed_the_World_in_2050.pdf.

Fleurence, J. and L. Levine. 2016. Seaweed in health and disease prevention, Academic Press. London, UK, 476 p.

Gorham, J. and S.A. Lewey. 1984. Seasonal-changes in the chemical-composition of *Sargassum muticum*. Marine Biology 80(1): 103–107.

Guedes, A.C., H.M. Amaro and F.X. Malcata. 2011. Microalgae as a source of high added-value compounds—a brief review of recent work. Biotechnology Progress 27: 597–613.

Hill, J., E. Nelson, E. Tilmand, S. Polasky and D. Tiffany. 2006. Environmental, economic and energetic costs and benefits of biodiesel and ethanol biofuels. PNAS 103: 11206–11210.

Huang, G.H., C. Feng, W. Dong, Z. XueWu and C. Gu. 2010. Biodiesel production by microbial biotechnology. Applied Energy 87: 38–46.

Jang, J.S., Y. Cho, G.T. Jeong and S.K. Kim. 2012. Optimization of saccharification and ethanol production by simultaneous saccharification and fermentation (SSF) from seaweed *Saccharina japonica*. Bioprocess Biosyst. Eng. 35: 11–18.

Khambhaty, Y., K. Mody, M.R. Gandhi, S. Thampy, P. Maiti and H. Brahmbhatt. 2012. *Kappaphycus alvarezii* as a source of bioethanol. Bioresource Technol. 103: 180–185.

Kraan, S. 2012. Algal polysaccharides, novel applications and outlook, pp. 489–532. *In:* Chuan-Fa Chang (ed.). Carbohydrates—Comprehensive Studies on Glycobiology and Glycotechnology. InTech. Available online at: http://www.intechopen.com/books/carbohydrates-comprehensive-studies-on-glycobiology-and-glycotechnology/algal-polysaccharides-novel-applications-and-outlook.

Kraan, S. 2013. Mass-cultivation of carbohydrate rich macroalgae, a possible solution for sustainable mitigation and adaptation strategies for global change. Biofuel Production 18(1): 27–46.

Lee, J., P. Li, J. Lee, H.J. Ryu and K.K. Oh. 2013. Ethanol production from *Saccharina japonica* using an optimized extremely low acid pretreatment followed by simultaneous saccharification and fermentation. Bioresour. Technol. 127: 119–125.

Liu, Z.L., P.J. Slininger, B.S. Dien, M.A. Berhow, C.P. Kurtzman and S.W. Gorsich. 2004. Adaptive response of yeasts to furfural and 5-hydroxymethylfurfural and new chemical evidence for HMF conversion to 2,5-bis-hydroxymethylfuran. J. Ind. Microbiol. Biotechnol. 31(8): 345–352.

Mata, T.M., A.A. Martins and N.S. Caetano. 2010. Microalgae for biodiesel production and other applications: A review. Renew. Sustain. Energy 14: 217–232.

Miller, R.G. and S.R. Sorrell. 2014. The future of oil supply. Phil. Trans. R. Soc. A. 372: 1–27.

Murphy, F., G. Devlin, R. Deverell and K. McDonnell. 2013. Biofuel production in Ireland—an approach to 2020 targets with a focus on algal biomass. Energies 6: 6391–6412.

Myra, G., A. Borines, R. de Leon and J. Cuello. 2013. A bioethanol production from the macroalgae *Sargassum*. Bioresource Technology 138: 22–29.

Nazaré, R. 2015. Produção de bioetanol a partir de *Sargassum muticum* (Phaeophyceae). MsC Thesis, University of Coimbra, Coimbra, Portugal.

Rodrigues, D., A.C. Freitas, L. Pereira, T.A. Rocha-Santos, M.W. Vasconcelos, M. Roriz et al. 2015. Chemical composition of red, brown and green macroalgae from Buarcos bay in Central West Coast of Portugal. Food Chem. 183: 197–207.

Sánchez-Machado, D.I., J. López-Cervantes, J. López-hernández, P. Paseiro-Losada and J. Simal-Lozano. 2004. Determination of the uronic acid composition of seaweed dietary fibre by HPLC. Biomed. Chromatogr. 18: 90–97.

Srivastava, N., R. Rawat, H. Oberoi and H. Singh. 2015. A review on fuel ethanol production from lignocellulosic biomass. International Journal of Green Energy 12(9): 949–960.

Trivedi, N., V. Gupta, C.R.K. Reddy and B. Jha 2015. Marine bioenergy: Trends and developments in Marine macroalgal biomass as a renewable source of bioethanol pp. 197–216. *In:* Se-kwon Kim and Choul-Gyun Lee (eds.). CRC Press, Taylor and Francis Group, Boca Raton, FL.

van Maris, A.J., D.A. Abbott, E. Bellissimi, J. van den Brink, M. Kuyper, M.A. Luttik, H. W. Wisselink, W.A. Scheffers, J.P. van Dijken and J.T. Pronk. 2006. Alcoholic fermentation of carbon sources in biomass hydrolysates by *Saccharomyces cerevisiae*: Current status. Antonie Van Leeuwenhoek 90(4): 391–418.

Wang, D.M., H.T. Kim, E.J. Yun, D.H. Kim, Y.C. Park, H.C. Woo and K.H. Kim. 2014. Optimal production of 4-deoxy-l-erythro-5-hexoseulose uronic acid from alginate for brown macro algae saccharification by combining endo- and exo-type alginate lyases. Bioprocess Biosyst. Eng. 37(10): 2105–2111.

Wei, N., J. Quarterman and J. Yong-Su. 2013. Marine macroalgae: An untapped resource for producing fuels and chemicals. Trends in Biotechnology 31(2): 70–77.

Yanagisawa, M., K. Nakamura, O. Ariga and K. Nakasaki. 2011. Production of high concentrations of bioethanol from seaweeds that contain easily hydrolyzable polysaccharides. Process Biochem. 46: 2111–2116.

Index

Printed and bound by CPI Group (UK) Ltd, Croydon, CR0 4YY

01/11/2024

01782624-0020